AF180464

Fibroblast Growth Factors
Biology and Clinical Application

FGF Biology and Therapeutics

Fibroblast Growth Factors
Biology and Clinical Application

FGF Biology and Therapeutics

Michael Simons

Yale University, USA

World Scientific

NEW JERSEY • LONDON • SINGAPORE • BEIJING • SHANGHAI • HONG KONG • TAIPEI • CHENNAI • TOKYO

Published by

World Scientific Publishing Co. Pte. Ltd.

5 Toh Tuck Link, Singapore 596224

USA office: 27 Warren Street, Suite 401-402, Hackensack, NJ 07601

UK office: 57 Shelton Street, Covent Garden, London WC2H 9HE

Library of Congress Cataloging-in-Publication Data
Names: Simons, Michael, MD
Title: Fibroblast growth factors : biology and clinical application :
FGF biology and therapeutics / by Michael Simons.
Description: New Jersey : World Scientific, 2016. |
Includes bibliographical references and index.
Identifiers: LCCN 2016035688 | ISBN 9789813143364 (hardcover : alk. paper)
Subjects: | MESH: Fibroblast Growth Factors
Classification: LCC QP552.F5 | NLM QU 107 | DDC 573.4--dc23
LC record available at https://lccn.loc.gov/2016035688

British Library Cataloguing-in-Publication Data
A catalogue record for this book is available from the British Library.

Copyright © 2017 by World Scientific Publishing Co. Pte. Ltd.

All rights reserved. This book, or parts thereof, may not be reproduced in any form or by any means, electronic or mechanical, including photocopying, recording or any information storage and retrieval system now known or to be invented, without written permission from the publisher.

For photocopying of material in this volume, please pay a copying fee through the Copyright Clearance Center, Inc., 222 Rosewood Drive, Danvers, MA 01923, USA. In this case permission to photocopy is not required from the publisher.

Typeset by Stallion Press
Email: enquiries@stallionpress.com

Printed in Singapore

Preface

Fibroblast growth factor family is the largest and most diverse group of growth factors involved in many aspects of normal biology and disease pathogenesis. FGFs have the additional distinction of being the first growth factors identified. Although FGF2 (basic FGF), the first FGF to be purified, was originally isolated from the bovine pituitary extract based on its ability to stimulate proliferation of BALB/c 3T3 fibroblasts, subsequent studies showed a very broad expression pattern. Similarly, FGF1 (acidic FGF), while initially isolated based on its ability to stimulate growth of endothelial cells, was subsequently shown to also act on a broad spectrum of cells. Subsequent studies using heparin sepharose chomatography and molecular cloning greatly expanded the size and diversity of the FGF family.

The FGF family of ligands and the corresponding family of FGF receptor tyrosine kinases comprise one of the most versatile and diverse growth factor signaling families in vertebrates, playing critical roles in a wide variety of biological processes. In addition to the twenty-two FGF ligands and four tyrosine kinase receptors, there is a number of co-receptors and cell-surface and cytoplasmic proteins that modulate FGF signaling in a tissue and cell-type specific manner. This great diversity and complexity of FGF biology is at the core of the multitude of biological activities and role ascribed to these growth factors.

Recent years have seen major expansion of knowledge about this family of proteins and it would be impossible to cover all aspects of their biology with any degree of completeness. For that reason this monograph is focused largely on vascular effects of FGFs. This choice is motivated, in part, by recent advances in our understanding of FGFs' role in the vasculature and by the growing understanding of vasculature's contribution to the regulation of normal organ homeostasis as well as better appreciation of the role of abnormal FGF singaling in a wide array of disease processes. With this focus in mind, we explore everything from the structure and signaling of FGFs and their receptors (Chapters 1, 2 and 3) to their role in the regulation of vascular homeostasis (Chapter 4) to their role in cardiovascular diseases (Chapters 5 and 6). We then address new developments in FGF therapeutics (Chapter 7) and explore FGF biology in the epithelium (Chapter 8) thereby providing a complete analysis of this growth factor family biology in two closely related but distinctly different environments- endothelium and epithelium. Finally, we examine FGF biology in eye disease and in cancer.

It is our hope that this comprehensive treatment of FGF vascular biology and its application will provide the reader with the accurate and timely summary of this rapidly moving field.

Contents

Chapter 1

An Introduction to the Fibroblast Growth Factors

David M. Ornitz*,‡ and Nobuyuki Itoh†

Abstract

Fibroblast growth factors (FGFs) comprise a family of twenty-two evolutionarily conserved molecules. Eighteen of the FGFs are secreted signaling molecules that interact with four tyrosine kinase FGF receptors (FGFRs), other FGF binding proteins, and heparan sulfate proteoglycans or Klotho family proteins. FGFs and FGFRs are present in nearly every tissue of a developing vertebrate and consequently a wide spectrum of developmental defects result from mutations in FGF ligands, receptors, interacting proteins, and downstream mediators and effectors of the signaling pathway. FGFs and FGFRs are also present in adult tissues where they regulate metabolic and physiologic function, maintain tissue homeostasis, and mediate injury response, tissue repair, and regeneration.

*Department of Developmental Biology, Washington University School of Medicine, St. Louis, MO 63110, USA, E-mail: dornitz@wustl.edu

†Medical Innovation Center, Kyoto University Graduate School of Medicne, Sakyo, Kyoto 606-8507, Japan

‡Corresponding author.

Mutation, gene amplification, and ectopic expression of FGFs and FGFRs can also lead to the initiation or progression of cancer. Four members of the FGF family are intracellular molecules that interact with and regulate the activity of voltage-gated sodium channels, to regulate cell excitability.

1. Introduction to the FGF Family

The mammalian FGF family contains twenty-two genes; eighteen encode secreted molecules that signal through tyrosine kinase FGF receptors (FGFRs) and four encode intracellular molecules (iFGFs) that regulate voltage-gated sodium (Nav) channels.[1,2] Based on biochemical function, sequence similarities, and evolutionary conservation, FGFs can be grouped into seven subfamilies. The secreted FGFs are categorized into five subfamilies of locally acting (autocrine or paracrine) FGFs and one subfamily of endocrine FGFs, and the intracellular FGFs are categorized into a separate subfamily (Fig. 1).[1–9] FGF15 in rodents, and FGF19 in other vertebrates, are thought to be orthologs and are referred to as FGF15/19.

2. Discovery of FGFs

The concept that a soluble factor that could promote the growth of tissues or cells was developed in the early 1900's using extracts from tissues or embryos to promote the growth of primary fibroblasts.[10–12] It was not until the 1970's that an extract from bovine pituitary was shown to stimulate the proliferation of 3T3 fibroblasts at nanogram per ml concentrations of protein. Protease sensitivity and thermolability suggested that this "Fibroblast growth factor" activity was contained within a protein,[13] partial purification of this activity was achieved in 1975 by Gospodarowicz,[14] and a homogeneous preparation was isolated by Lemmon in 1983.[15] This protein was referred to as basic FGF (bFGF or FGF2) due the overall basic composition of amino acids and high isoelectric point. A second Fibroblast growth

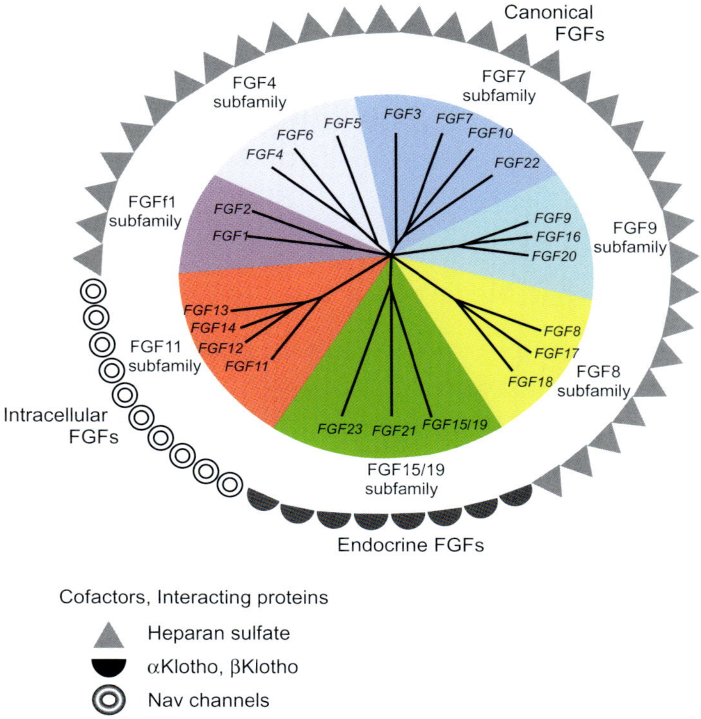

Fig. 1. Phylogenetic analysis segregates the twenty-two FGF genes into seven subfamilies that contain two to four members each. Branch lengths are proportional to the evolutionary distance between each gene. Canonical FGFs, which bind to and activate FGFRs with heparin/HS as a co-factor, include the FGFs 1, 4, 7, 8, and 9 subfamilies. Endocrine FGFs include the FGF15/19 subfamily. These FGFs bind and activate FGFRs but use Klotho family proteins as a co-factor. Intracellular FGFs (iFGFs) are also referred to as fibroblast growth factor homologous factors (FHFs) and include the FGF11 subfamily. These are non-signaling proteins that regulate Nav channels and other cellular proteins.

factor-like activity with a low isoelectric point was purified from bovine brain. This factor was referred to as acidic FGF (aFGF or FGF1).[16-21] FGF1 turned out to be identical to a factor called ECGF (endothelial cell growth factor), which was consistent with its potent mitogenic activity towards endothelial cells.[15,18,22–24] FGF1 was first cloned from a human brain cDNA library[25] and FGFs 1 and 2 were cloned from bovine pituitary cDNA libraries[26] in 1986. The remaining members of the FGF family were identified based on oncogenic properties activated by retroviral insertions, as growth factors for cultured cells or tissues, as genes that were mutated in genetic diseases, by homology-based cloning, or homology-based database searches.[1,3,27,28]

3. Molecules that Mediate the Function of FGFs

Secreted FGFs transmit their signal to cellular targets through four FGFRs (**detailed in chapter 2**) (reviewed in Ornitz and Itoh, 2015).[1] The ligand binding affinity and specificity of FGFRs is regulated through alternative splicing of the FGFR mRNAs, and interactions with co-factor molecules. Several alternative splicing events occur in various extracellular and intracellular encoding regions; however, alternative splicing in the third immunoglobulin-like domain of FGFRs 1-3 generating b and c variants are particularly important determinants of ligand binding specificity. Interestingly, FGF1 is the only ligand that can activate all FGFRs irrespective of alternative splicing.[29] Ligand binding specificity assays show that the FGF7 subfamily preferentially binds and activates b splice variants while the other subfamilies activate c splice variants. The FGF9 subfamily also uniquely activates FGFR3b.[29,30] For the canonical FGF subfamilies, (FGF1, 4, 7, 8, and 9 subfamilies) heparan sulfate (HS) serves as the primary endogenous co-factor for ligand binding and receptor activation (Fig. 1).[31–33] For endocrine FGFs (FGF15/19 subfamily), αKlotho and βKlotho function as co-factors.[34] Intracellular FGFs (FGF11 subfamily) directly bind and regulate Nav channels and do not activate FGFRs.[35,36].

4. The FGF Subfamilies

4.1. *FGF1 subfamily*

FGFs 1 and 2 were the first identified FGFs and now comprise the FGF1 subfamily (Fig. 1). FGFs 1 and 2 lack secretory signal peptides but are nevertheless found on the cell surface and in the extracellular matrix (ECM). The mechanism of export involves the direct translocation across the cell membrane mediated by a chaperone complex that includes synaptotagmin-1, the calcium binding protein S100A13, and phosphorylation of FGF2 by Tec kinase.[37–41] Alternatively, through plasma membrane disruptions, FGF2 can be directly released from cytoplasmic stores following cell injury.[42–44] FGFs 1 and 2 have also been found in the nucleus of some cells. Several studies have shown that extracellular FGF1 transits across the plasma membrane and cytosol before entering the nucleus.[45,46] The mechanisms by which FGFs move through the cell are thought to require binding and activation of cell surface FGFRs and interaction with the chaperone, HSP90.[47,48] The functions of nuclear FGF1 include regulation of the cell cycle, differentiation, and survival.[49,50]

FGFs 1 and 2 have relatively minor roles in embryonic development but are key factors involved in wound healing and cardioprotection in response to ischemic injury (Table 1).[51–58] FGF1 may also be an important metabolic regulator functioning in adipose tissue where its expression is induced in response to a high fat diet. Interestingly, mice lacking FGF1 develop a diabetes phenotype when placed on a high fat diet.[59] Pharmacological administration of FGF1 to diabetic mice signals predominantly through FGFR1 to increase insulin-dependent glucose uptake in skeletal muscle and reduce hepatic production of glucose.[60] Mice lacking FGF2 do not have major developmental defects, but as adults show reduced vascular tone, impaired cardiac hypertrophy, reduced cortical neuron density, and defects in response to cutaneous, pulmonary, or cardiac injury.[51,52,55,58,61–64]

4.2. *FGF4 subfamily*

The FGF4 subfamily is comprised of FGFs 4, 5 and 6 (Fig. 1).[2] However, based on synteny relationships FGF5 could alternatively be placed in the FGF1 subfamily.[6] FGFs 4, 5 and 6 are secreted proteins with cleavable N-terminal signal peptides. These FGFs mediate biological responses as extracellular proteins by binding to and activating c splice variants of FGFRs 1-3 and FGFR4.[29,30]

Table 1. Phenotypes of germline and conditional loss-of-function FGF mutations in mice.*

Gene name	Viability/ age at death of null mutant	Null phenotype (organ, structure, or cell type affected)	Tissue-specific (conditional) phenotypes, Redundant phenotypes, Phenotypes induced by physiological challenge	Selected references
FGF1	Viable	No apparent phenotype	An aggressive diabetic phenotype with white adipocyte remodelling on high-fat diet	51, 59
FGF2	Viable	Cortical neuron, vascular smooth muscle, blood pressure, skeletal development, and wound healing	Decreased cardiac hypertrophy induced by ischemic injury and delayed wound healing; Increased bone mineralization in high molecular weight isoform knockout	52, 53, 63, 229–232
FGF3	Viable	Inner ear and skeletal development	Heart development (redundant with FGF10)	80, 233
FGF4	E4-5	Blastocyst inner cell mass	Limb bud development (redundant with FGF8)	65–67, 234
FGF5	Viable	Hair follicle development		70
FGF6	Viable	Muscle development	Muscle regeneration	76, 79, 235
FGF7	Viable	Hair follicle and ureteric bud development and synaptogenesis	Thymus regeneration (radiation injury) and wound healing	87–89, 91, 236, 237
FGF8	E7	Gastrulation	Heart field, limb, somitogenesis, kidney, CNS, inner ear development, spermatogenesis	68, 106–108, 238–246

(*Continued*)

FGF9	P0	Lung, heart, skeletal, gonad, inner ear, and intestine development	Migration of cerebellar granule neurons, proliferation of mesenchyme associated with cochlea, and kidney agenesis (redundant with FGF20)	121–126, 128, 135, 136, 247–252
FGF10	P0	Limb bud, lung bud, trachea, thymus, pancreas, pituitary, palate, tongue epithelium, semicircular canal, cochlea, cecum, kidney, submandibular, salivary, lacrimal, and mammary gland, heart, stomach, and white adipose tissue	Lung branching morphogenesis and inner ear development (redundant with FGF3)	81, 94–96, 253–265
FGF11	Viable	No identified phenotype		(unpublished)
FGF12	Viable	No identified phenotype	Severe ataxia and motor weakness (redundant with FGF14)	199
FGF13	Viable	Neuronal migration, learning and memory deficits, synaptic transmission, and microtubule binding		199, 204, 206, 207, 214, 266
FGF14	Viable	Ataxia, motor weakness, learning and memory deficits, and impaired neuronal excitability	Severe ataxia and motor weakness (redundant with FGF12)	199, 205, 215
FGF15	E13.5-P7	Cardiac outflow tract development, neurogenesis, and bile acid metabolism, hepatocellular proliferation	Liver regeneration	154, 171, 172, 175, 267–269
FGF16	Viable	Heart development	Promotes cardiac hypertrophy and fibrosis induced by angiotensin II	129–131
FGF17	Viable	Cerebellum and frontal cortex development		108, 109
FGF18	P0	CNS, skeletal, palate, and lung development		110–113, 135, 270

Table 1. (*Continued*)

Gene name	Viability/ age at death of null mutant	Null phenotype (organ, structure, or cell type affected)	Tissue-specific (conditional) phenotypes, Redundant phenotypes, Phenotypes induced by physiological challenge	Selected references
FGF20	Viable	Guards hair, teeth, cochlea, and kidney development	Kidney agenesis and proliferation of mesenchyme associated with cochlea (redundant with FGF9)	132–136
FGF21	Viable	Energy/lipid metabolism	Cardioprotection following oxidative stress	154, 181, 271, 272
FGF22	Viable	Synaptogenesis	Decreased skin papillomas formation following carcinogenesis challenge	87, 91, 103, 104, 273, 274
FGF23	PW4-13	Phosphate and vitamin D homeostasis, deafness, middle ear development	Bone mineralization	27, 187, 274–278

*Adapted from Ornitz and Itoh, 2015.[1]

FGF4 knockout embryos degenerate shortly after implantation (Table 1). The etiology of this phenotype includes impaired proliferation and reduced survival of the inner cell mass.[65] In limb bud development, FGF4 is prominently expressed in the apical ectodermal ridge (AER) along with FGFs 8, 9 and 17. However, embryos with conditional inactivation of FGF4 in the AER had normal limb bud development.[66,67] Inactivating multiple FGFs simultaneously showed that FGF4 and FGF8 are functionally redundant and together are required for survival of dorsal proximal limb mesenchyme.[68] FGFs 4 and 8 also show redundancy in somitogenesis. Single conditional knockouts in presomitic mesoderm have normal somitogenesis, but inactivation of both genes results in loss and premature differentiation of presomitic mesoderm. This phenotype is rescued with a single wild type allele of either FGF4 or FGF8.[69] FGF5 and FGF6 knockout mice are viable (Table 1). Inactivation of the FGF5 gene in mice results in the angora phenotype (abnormally long hair).[70] In dogs, genome-wide association studies identified a mutation in FGF5 that is associated with hair length (Table 2).[71] A missense mutation in FGF5 was also found in longhaired cats, rabbits, and the Angora mouse mutant.[70,72,73] In humans, a mutation in FGF5 is also associated with abnormally long eyelash growth.[74] FGF6 is highly expressed in embryonic skeletal muscle and is induced following muscle injury in the adult.[75,76] A role for FGF6 in muscle regeneration has been controversial, but the consensus of several studies suggest that mice lacking FGF6 in muscle have delayed regeneration following injury.[77–79]

4.3. FGF7 *subfamily*

FGFs 3, 7, 10 and 22 are grouped together as the FGF7 subfamily based on phylogenetic analysis sequence homology and biochemical properties (Fig. 1).[2,30] FGFs, 3, 7, 10, and 22 preferentially activate the b splice variant of FGFR2. FGF3 and FGF10 also activate the b splice variant of FGFR1.[29,30] Even though biochemical and sequence similarities place FGF3 in the FGF7 subfamily, genetic linkage of FGF3 and FGF4 could support inclusion of FGF3 in the

Table 2. Heritable mutations in FGFs associated with disease in humans and other mammals.*

Gene name	Mutation	Associated disease	Selected references
FGF1			
FGF2			
FGF3	Haploinsufficiency	Oto-dental syndrome	82, 83, 85, 86, 279
	Missense/frameshift mutation	Michel aplasia (inner ear agenesis, microtia, and microdontia), labyrinthine aplasia, microtia, and microdontia (LAMM syndrome)	
FGF4	Retroviral overexpression	Chondrodysplasia (dogs)	234
FGF5	Deletion mutation	Angora mutation (mice)	70–72, 280, 281
	Missense/splice-site mutation	Coat variability (pure bred dogs)	
	Missense/insertion/deletion mutation	Long-hair (cats)	
FGF6			
FGF7	Polymorphism	Chronic obstructive pulmonary disease risk	93
FGF8	Nonsense mutation	Hypogonadotropic hypogonadism	115, 117, 282–285
	Missense mutation	Cleft lip and palate, Holoprosencephaly, craniofacial defects, Hypothalamo-pituitary dysfunction, Kallman syndrome type 6	
	Hypomorphic allele	Lack of hypothalamic GnRH neurons	
FGF9	Missense mutation	Multiple synostoses syndrome, Elbow knee synostosis (mice)	137–139
	Promoter polymorphism	Sertoli cell-only syndrome	
FGF10	Nonsense mutation	Aplasia of lacrimal and salivary glands, Lacrimo-auriculo-dento-digital (LADD) syndrome	98–101, 286
	Polymorphism	Extreme myopia	

FGF11			
FGF12	Missense mutation	Brugada syndrome (candidate gene) Kashin-Beck disease (candidate gene)	220, 287
FGF13	Nonsense mutation Position effect Translocation	Börjeson-Forssman-Lehmann syndrome (BFLS) (candidate gene) X-linked congenital generalized hypertrichosis Genetic Epilepsy and Febrile Seizures Plus (GEFS+)	221, 222, 266
FGF14	Missense mutation/ translocation/deletion	Spinocerebellar ataxia 27 (SCA27) Autosomal dominant episodic ataxia	223, 225, 288, 289
FGF15/19			
FGF16	Nonsense mutation	Metacarpal 4–5 fusion	141, 142
FGF17	Missense mutation	Hypogonadotropic hypogonadism	116
FGF18	Polymorphism	Nonsyndromic cleft lip and palate	117
FGF20	Polymorphism Missense mutation	Parkinson disease risk Kidney agenesis (human)	135, 144–147
FGF21	Polymorphism	Macronutrient intake, obesity, and type-2 diabetes risk	183–185
FGF22			
FGF23	Missense mutation	Autosomal dominant hypophosphataemic rickets (ADHR), Hyperphosphatemic familial tumoral calcinosis (HFTC)	155, 191–194, 290–292
	Polymorphism	Cardiac abnormality risk in Kawasaki disease (increased serum FGF23)	

*Adapted from Ornitz and Itoh, 2015.[1]

FGF4 subfamily instead of the FGF7 subfamily.[6] Genomic sequence analysis of chordates suggests that FGF3 could alternatively be considered as an individual member of its own subfamily.[5]

FGF3 knockout mice are viable, but have phenotypes that include inner ear agenesis and dysgenesis, microtia, and microdontia (Table 1).[80–82] Autosomal recessive mutations are found in FGF3 in humans (Table 2). These result in inner ear labyrinth malformations, microtia, and microdontia (LAMM) syndrome and related Michel aplasia, characterized by type I microtia, microdontia, and profound congenital deafness associated with a complete absence of inner ear structures.[83–86]

FGF7 knockout mice are viable but have impaired hair and kidney development and defects in the formation of inhibitory neuronal synapses.[87–91] Mice lacking FGF7 also have impaired liver regeneration due to inability to expand progenitor cell populations.[92] In humans, genome wide association studies identified single nucleotide polymorphisms in FGF7 significantly associated with risk for chronic obstructive pulmonary disease (COPD).[93]

FGF10 knockout mice lack lungs and limbs, and consequently die shortly after birth. FGF10 is expressed in mesenchymal cells and signals to nearby epithelium. FGF10 regulates epithelial-mesenchymal interactions that are necessary for the development of epithelial components of multiple organs including the limb, lung, salivary glands kidney, and white adipose tissue.[94–97] Two autosomal dominant genetic diseases, Aplasia of the lacrimal and salivary glands (ALSG) and Lacrimo-auriculo-dento-digital (LADD) syndrome, are associated with mutations in FGF10.[98–100] ALSG is characterized by aplasia, atresia, or hypoplasia of the lacrimal and salivary systems. LADD syndrome is characterized by aplasia, atresia, or hypoplasia of the lacrimal and salivary systems, cup-shaped ears, hearing loss, and dental and digital anomalies. Severe myopia (nearsightedness) is associated with a single nucleotide polymorphism in FGF10.[101] In strong support of an FGF10-FGFR2b signal, loss-of-function mutations in FGFR2 are also a cause of LADD syndrome.

FGF22 knockout mice are viable, but have defects in synaptogenesis.[102] Interestingly, in the absence of FGF22 the formation of excitatory synapses is impaired. This is in contrast to defects in inhibitory synaptogenesis in FGF7 knockout mice. Phenotypically, mice lacking FGF7 and FGF22 are either resistant to or prone to epileptic seizures, respectively.[87,103] Recent studies show that FGF22 functions as an organizing molecule in the dorsal lateral geniculate nucleus.[104] The mechanism by which FGF7 localizes to postsynaptic sites for inhibitory synapses and FGF22 localized to postsynaptic sites for excitatory synapses involves selective axonal transport mediated by motor proteins KIF5, and KIF3A and KIF17, respectively.[91]

4.4. *FGF8 subfamily*

FGF8, FGF17, and FGF18 comprise the FGF8 subfamily (Fig. 1).[2] These FGFs are classically secreted molecules that activate c splice variants of FGFRs 1–3 and FGFR4.[29,30]

Mice lacking FGF8 die by embryonic day 9.5 and are missing all embryonic mesoderm and endoderm-derived structures (Table 1).[105] FGF8 signaling is also required for FGF4 expression in the primitive streak, thus the impaired migration away from the primitive streak could also reflect redundant function of FGF4 and FGF8.[106] Inactivation of FGF8 in the limb bud apical ectodermal ridge identified additional roles in limb bud development.[107] FGF17 knockout mice are viable, but loss of expression of FGF17 in the mid-hindbrain junction results in loss of the anterior lobe of the cerebellar vermis and a selective reduction in the size of the dorsal frontal cortex.[108,109] FGF18 knockout mice die in the perinatal period, probably due to cleft palate. Mice lacking FGF18 have impaired development of mesenchymal components of the skeleton, lung, and brain.[110–114] FGF18 also has a direct role in regulating alveolar development at late gestational stages.[113]

In humans, mutations in FGF8 cause familial hypogonadotropic hypogonadism, and depending on the mutation, variable deficiency of gonadotropin-releasing hormone and olfactory phenotypes characteristic of Kallmann syndrome (Table 2).[115] Mutations in FGF17 were also found in three patients with familial hypogonadotropic hypogonadism, two of which were classified as Kallmann syndrome.[116]

A predicted loss-of-function missense mutation in FGF8 was also found in a patient with cleft lip and palate.[117]

4.5. *FGF9 subfamily*

FGF9, FGF16, and FGF20 comprise the FGF9 subfamily (Fig. 1). This subfamily of FGFs is secreted from cells. Although these FGFs lack a classical N-terminal signal peptide, a non-cleaved internal hydrophobic sequence functions as a signal for transport into the endoplasmic reticulum and secretion from cells.[118–120] The FGF9 subfamily uniquely activates the b splice variant of FGFR3 in addition to c splice variants of FGFRs 1–3 and FGFR4.[29,30]

FGF9 knockout mice have organogenesis defects that affect lung, gonad, skeletal, heart, kidney, small intestine and cecum, and inner ear development (Table 1).[121–128] Mice lacking FGF16 are viable but have decreased cardiomyocyte proliferation in embryos and neonatal stages,[129,130] and enhanced cardiac hypertrophy and fibrosis in response to angiotensin II as adults.[131] Mice lacking FGF20 are viable but have developmental defects that affect hair, cochlea, kidney, and tooth development.[132–135] FGF9 and FGF20 show redundancy in kidney and inner ear development. In the developing kidney, FGF9 and FGF20 maintain the stemness of cap mesenchyme progenitor cells and in inner ear development they regulate sensory progenitor proliferation indirectly through signaling to adjacent mesenchyme.[136,136]

In humans, an autosomal dominant missense mutation in FGF9 impairs receptor binding, reduces chondrocyte proliferation, and increases osteoblast differentiation and matrix mineralization, resulting in multiple synostosis syndrome (SYNS) (Table 2).[137] A similar missense mutation in FGF9 in mice causes elbow knee synostosis (EKS) syndrome.[138] A gene regulatory mutation in the FGF9 promoter causes decreased expression of FGF9 in Lydig cells, resulting in the Sertoli cell–only syndrome (SCOS), in which affected patients have atrophic testes, azoospermia, and hypogonadism.[139] Nonsense mutations have been identified in FGF16 associated with X-linked recessive metacarpal 4–5 fusion and fifth digit hypoplasia (MF4).[140–143]

A frameshift mutation has been identified in FGF20 in a family with bilateral renal agenesis.[135] Single nucleotide polymorphisms, one of which is in the microRNA-433 binding site in the 3′ UTR of FGF20, have been associated with Parkinsons' disease in some, but not all, studies. This finding suggests that altered expression of FGF20 may be a risk factor for Parkinsons' disease.[144–153]

4.6. *FGF15/19 subfamily* (*endocrine FGFs*)

FGF15/19, FGF21, and FGF23 comprise the FGF15/19 subfamily (Fig. 1).[154,155] The FGF15/19 subfamily primarily functions as endocrine factors (eFGFs). eFGFs bind to heparin/HS with very low affinity, which facilitates release from the ECM and allows these FGFs to function as endocrine factors.[156] eFGFs mediate their biological responses through FGFRs, but require members of the Klotho family [αKlotho (Klotho), βKlotho, Klotho-LPH related protein (KLPH, also called Lactase-like Klotho or γKlotho)] as co-factors for receptor binding and activation. FGF23 requires αKlotho, and FGF15/19 and FGF21 require βKlotho[154,157–161] KLPH can enhance FGF15/19 signaling in HEK293 cells,[162] however, the *in vivo* function of KLPH is not known. *In vitro* receptor activation assays show that FGF15/19 activates c splice variants of FGFRs 1–3 and FGFR4, while FGF21 only activates the c splice variants of FGFRs 1 and 3.[161,163] FGF23 signals through c splice variants of FGFRs 1 and 3, and FGFR4, together with αKlotho.[164–168]

FGF15/19 specifically activates FGFR4 in hepatocytes to induce proliferation and to regulate bile acid synthesis.[163,169,170] In postnatal animals, intestinal-derived FGF15/19 regulates hepatic bile acid synthesis.[171] Mice lacking FGF15 gradually die due to variably penetrant defects in development of the cardiac outflow tract (Table 1).[172,173] These mice also have severe defects in liver regeneration following partial hepatectomy.[174,175] FGF19 and FGFR4 signaling also contribute to the etiology or progression of hepatocellular carcinoma.[176]

FGF21 directly regulates hepatocyte and adipocyte metabolism through interactions with FGFR1 and βKlotho.[160,163,177] When challenged by fasting, FGF21 expression is rapidly upregulated in the liver.[178–180] FGF21 functions to induce lipolysis in white adipose tissue and ketogenesis in the liver.[179] Mice lacking FGF21 do not show overt phenotypes under homeostatic conditions; however, they do show adipocyte hypertrophy, decreased lipolysis, and decreased serum nonesterified fatty acid (NENF) levels.[181] They also show impaired adaptation to a ketogenic diet.[182] In response to fasting, mice lacking FGF21 showed increased adipocytes lipolysis and increased serum NENF levels.[181] In humans, polymorphisms in FGF21 are potentially associated with increased risk of obesity and type 2 diabetes (Table 2).[183–185]

FGF23 is expressed in osteocytes and signals to the kidney where it functions to inhibit renal phosphate reabsorption.[186] FGF23 knockout mice survive until birth, but show severe growth retardation, abnormal bones, and a shortened life span. These mice have high serum phosphate levels due to increased renal phosphate reabsorption, and elevated serum 1,25$(OH)_2$ vitamin D caused by increased expression of renal 25-hydroxyvitamin D3-1α–hydroxylase (Cyp27B1).[187] FGF23 also signals directly to cardiomyocytes to induce hypertrophy,[188] and increase myocyte Ca^{2+} levels and cardiac contractility.[189]

Gain-of-function mutations in FGF23 result in autosomal dominant hypophosphatemic rickets (AHDR) in humans.[155] Tumors that over-produce FGF23 cause tumor-induced osteomalacia, a paraneoplastic disease characterized by renal phosphate wasting and hypophosphatemia.[190] Reduced FGF23 signaling causes familial tumoral calcinosis characterized by ectopic calcification and hyperphosphatemia.[191,192] Polymorphisms in FGF23 and elevated serum levels of FGF23 are associated with Kawasaki syndrome (KS), a childhood vascular inflammatory disease with an increased risk of developing subsequent cardiac abnormalities.[193,194]

4.7. *FGF11 subfamily*

FGFs 11–14 comprise the FGF11 subfamily and are also known as fibroblast growth factor homologous factors (FHFs) and intracellular FGFs (iFGFs, Fig. 1).[35,195,196] iFGFs are not secreted and have no known interaction with signaling FGFRs.[36] iFGFs bind to the cytosolic carboxy terminal tail of Nav channels and regulate their subcellular localization and ion-gating properties in mature neurons and in other excitable cells such as cardiomyocytes.[197–202] Additional interacting proteins have been identified for some iFGFs and include the mitogen-activated protein kinase (MAPK) scaffolding protein, IB2 (MAPK8IP2),[203] and microtubules.[204]

Mice lacking FGF13 are viable but have impaired neuronal migration and deficits in learning and memory (Table 1).[204] Mice lacking FGF14 have paroxysmal dyskinesia, movement disorders, and impaired spatial learning.[205–207] FGF14 and other members of the iFGF family interact with the cytoplasmic carboxy terminal tail of Nav channel α subunits.[199–201,208–211] FGF13 was also found to interact directly with and stabilize microtubules[204] and bind junctophilin-2, a protein that regulates L-type Ca^{2+} channels.[212]

Mice lacking FGF14 have defective neuronal firing due to altered Nav channel physiology (Table 2A).[199,207,213,214] Inactivation of FGF14 in adult mouse Purkinje neurons results in loss of spontaneous firing and deficits in coordination,[215] demonstrating that FGF14 is a physiological regulator of Nav channels *in vivo*. Furthermore, FGF14 interactions with Nav channels may be regulated by glycogen synthase kinase 3 which links cellular signaling pathways and neuronal excitability.[211,216–219]

In humans, a missense mutation in FGF12 has been identified in a patient with Brugada syndrome (inherited cardiac arrhythmia) (Table 2).[220] A duplication breakpoint was identified in a patient with Börjeson-Forssman-Lehmann syndrome (X-linked mental retardation) that maps near the FGF13 (FHF2) gene.[221] X-linked congenital generalized hypertrichosis (hair overgrowth) also maps near FGF13 and FGF13 expression is decreased in the outer root sheath of affected hair follicles, suggesting a role for FGF13 in hair follicle growth.[222] Mutations in FGF14 result in a progressive spinocerebellar ataxia syndrome (SCA27).[223–225] SCA27 is caused by missense, translocation, or deletion mutations in FGF14.[224,226,227] Loss of binding of the mutant FGF14 protein to Nav channel α subunits and instability of the mutant protein are thought to be the primary factors leading to this disease.[223,228]

5. Conclusions and Future Directions

Canonical and endocrine FGFs encode a large family of related proteins that regulate developmental, physiological, and pathological processes, including the earliest stages of embryonic development, organogenesis, tissue maturation, tissue homeostasis, metabolism, response to injury, and cancer. Mechanisms have evolved to regulate the expression FGFs and their paracrine and endocrine bioavailability. Other regulatory mechanisms have evolved to control the ability of FGFs to activate FGFRs and downstream cellular signaling.

In vivo studies are just beginning to identify additional redundant functions of FGFs, and how they interact with other intercellular signaling pathways. The discovery of endocrine FGFs has opened new and unanticipated mechanisms to regulate metabolism, lipid, and mineral homeostasis. Future studies aimed at understanding pathogenic mechanisms resulting from genetic and acquired alterations in FGF and FGFR signaling will provide new therapeutic opportunities to prevent or treat both inherited and acquired disease. For tissue repair and regeneration, it will be important to understand how FGF signaling pathways become reactivated and then suppressed in response to injury. Due to the diversity of FGFs and their ability to signal through multiple FGFRs, the development of ligand- and receptor-specific pharmacological agonists and antagonists will be required to achieve disease-specific efficacy and limit unwanted side effects.

References

1. Ornitz, D.M. and Itoh, N. The Fibroblast Growth Factor signaling pathway. *Wiley Interdiscip Rev Dev Biol* **4**, 215–266 (2015).
2. Itoh, N. and Ornitz, D.M. Evolution of the FGF and FGFr gene families. *Trends Genet* **20**, 563–569 (2004).
3. Beenken, A. and Mohammadi, M. The FGF family: biology, pathophysiology and therapy. *Nat Rev Drug Discov* **8**, 235–253 (2009).
4. Krejci, P., Prochazkova, J., Bryja, V., Kozubik, A. and Wilcox, W.R. Molecular pathology of the fibroblast growth factor family. *Human Mutat* **30**, 1245–1255 (2009).

5. Oulion, S., Bertrand, S. and Escriva, H. Evolution of the FGF Gene Family. *Int J Evol Biol* **2012**, 298147 (2012).
6. Itoh, N. and Ornitz, D.M. Functional Evolutionary History of the Mouse FGF Gene Family. *Dev Dynam* **237**, 18–27 (2008).
7. Itoh, N. and Ornitz, D.M. Fibroblast growth factors: From molecular evolution to roles in development, metabolism and disease. *J Biochem* **149**, 121–130 (2011).
8. Ornitz, D.M. and Itoh, N. Fibroblast growth factors. *Genome Biol* **2**, REVIEWS3005 (2001).
9. Itoh, N. Hormone-like (endocrine) FGFs: their evolutionary history and roles in development, metabolism, and disease. *Cell Tissue Res* **342**, 1–11 (2010).
10. Trowell, O.A., Chir, B. and Willmer, E.N. Studies on the growth of tissues *in vitro*. *J Exp Biol* **16**, 60–70 (1939).
11. Carrel, A. Artificial activation of the growth *in vitro* of connective tissue. *J Exp Med* **17**, 14–19 (1913).
12. Carrel, A. Contributions to the study of the mechanism of the growth of connective tissue. *J Exp Med* **18**, 287–298 (1913).
13. Armelin, H.A. Pituitary extracts and steroid hormones in the control of 3T3 cell growth. *Proc Natl Acad Sci USA* **70**, 2702–2706 (1973).
14. Gospodarowicz, D. Purification of a fibroblast growth factor from bovine pituitary. *J Biol Chem* **250**, 2515–2520 (1975).
15. Lemmon, S.K. and Bradshaw, R.A. Purification and partial characterization of bovine pituitary fibroblast growth factor. *J Cell Biochem* **21**, 195–208 (1983).
16. Thomas, K.A., Rios-Candelore, M. and Fitzpatrick, S. Purification and characterization of acidic fibroblast growth factor from bovine brain. *Proc Natl Acad Sci USA* **81**, 357–361 (1984).
17. Thomas, K.A., Riley, M.C., Lemmon, S.K., Baglan, N.C., *et al.* Brain fibroblast growth factor: Nonidentity with myelin basic protein fragments. *J Biol Chem* **255**, 5517–5520 (1980).
18. Lemmon, S.K., *et al.* Bovine fibroblast growth factor: comparison of brain and pituitary preparations. *J Cell Biol* **95**, 162–169 (1982).
19. Gimenez-Gallego, G., Conn, G., Hatcher, V.B. and Thomas, K.A. The complete amino acid sequence of human brain-derived acidic fibroblast growth factor. *Biochem Biophys Res Commun* **138**, 611–617 (1986).
20. Gimenez-Gallego, G., *et al.* Brain-derived acidic fibroblast growth factor: complete amino acid sequence and homologies. *Science* **230**, 1385–1388 (1985).

21. Thomas, K.A., *et al.* Pure brain-derived acidic fibroblast growth factor is a potent angiogenic vascular endothelial cell mitogen with sequence homology to interleukin 1. *Proc Natl Acad Sci USA* **82**, 6409–6413 (1985).
22. Gambarini, A.G. and Armelin, H.A. Purification and partial characterization of an acidic fibroblast growth factor from bovine pituitary. *J Biol Chem* **257**, 9692–9697 (1982).
23. Schreiber, A.B., *et al.* A unique family of endothelial cell polypeptide mitogens: the antigenic and receptor cross-reactivity of bovine endothelial cell growth factor, Brain-derived acidic fibroblast growth factor, and eye-derived growth factor-II. *J Cell Biol* **101**, 1623–1626 (1985).
24. Schreiber, A.B., *et al.* Interaction of endothelial cell growth factor with heparin: Characterization by receptor and antibody recognition. *Proc Natl Acad Sci USA* **82**, 6138–6142 (1985).
25. Jaye, M., *et al.* Human endothelial cell growth factor: Cloning, nucleotide sequence, and chromosome localization. *Science* **233**, 541–545 (1986).
26. Abraham, J.A., *et al.* Nucleotide sequence of a bovine clone encoding the angiogenic protein, basic fibroblast growth factor. *Science* **233**, 545–548 (1986).
27. Yu, X. and White, K.E. FGF23 and disorders of phosphate homeostasis. *Cytokine Growth Factor Rev* **16**, 221–232 (2005).
28. Turner, N. and Grose, R. Fibroblast growth factor signalling: From development to cancer. *Nat Rev Cancer* **10**, 116–129 (2010).
29. Ornitz, D.M., *et al.* Receptor specificity of the fibroblast growth factor family. *J Biol Chem* **271**, 15292–15297 (1996).
30. Zhang, X., *et al.* Receptor specificity of the fibroblast growth factor family. The complete mammalian FGF family. *J Biol Chem* **281**, 15694–15700 (2006).
31. Ornitz, D.M. FGFs, heparan sulfate and FGFRs: complex interactions essential for development. *Bioessays* **22**, 108–112 (2000).
32. Rapraeger, A.C., Krufka, A. and Olwin, B.B. Requirement of heparan sulfate for bFGF-mediated fibroblast growth and myoblast differentiation. *Science* **252**, 1705–1708 (1991).
33. Yayon, A., Klagsbrun, M., Esko, J.D., Leder, P., *et al.* Cell surface, heparin-like molecules are required for binding of basic fibroblast growth factor to its high affinity receptor. *Cell* **64**, 841–848 (1991).
34. Kuro-o, M. Endocrine FGFs and Klothos: Emerging concepts. *Trends Endocrinol Metab* **19**, 239–245 (2008).

35. Goldfarb, M. Fibroblast growth factor homologous factors: Evolution, structure, and function. *Cytokine Growth Factor Rev* **16**, 215–220 (2005).
36. Olsen, S.K., *et al.* Fibroblast growth factor (FGF) homologous factors share structural but not functional homology with FGFs. *J Biol Chem* **278**, 34226–34236 (2003).
37. Prudovsky, I., *et al.* The non-classical export routes: FGF1 and IL-1alpha point the way. *J Cell Sci* **116**, 4871–4881 (2003).
38. Landriscina, M., *et al.* S100A13 participates in the release of fibroblast growth factor 1 in response to heat shock *in vitro*. *J Biol Chem* **276**, 22544–22552 (2001).
39. Prudovsky, I., Kumar, T.K., Sterling, S. and Neivandt, D. Protein-phospholipid interactions in nonclassical protein secretion: Problem and methods of study. *Int J Mol Sci* **14**, 3734–3772 (2013).
40. Lavallee, T.M., *et al.* Synaptotagmin-1 Is Required For Fibroblast Growth Factor-1 Release. *J Biol Chem* **273**, 22217–22223 (1998).
41. La Venuta, G. et al. Small molecule inhibitors targeting Tec kinase block unconventional secretion of fibroblast growth factor 2. *J Biol Chem*, doi:10.1074/jbc.M116.729384 (2016).
42. Aizman, I., Vinodkumar, D., McGrogan, M. and Bates, D. Cell Injury-Induced Release of Fibroblast Growth Factor 2: Relevance to Intracerebral Mesenchymal Stromal Cell Transplantations. *Stem Cells Dev* **24**, 1623–1634 (2015).
43. McNeil, P.L., Muthukrishnan, L., Warder, E. and D'Amore, P.A. Growth factors are released by mechanically wounded endothelial cells. *J Cell Biol* **109**, 811–822 (1989).
44. Muthukrishnan, L., Warder, E. and McNeil, P.L. Basic fibroblast growth factor is efficiently released from a cytolsolic storage site through plasma membrane disruptions of endothelial cells. *J Cell Physiol* **148**, 1–16 (1991).
45. Olsnes, S., Klingenberg, O. and Wiedlocha, A. Transport of exogenous growth factors and cytokines to the cytosol and to the nucleus. *Physiol Rev* **83**, 163–182 (2003).
46. Planque, N. Nuclear trafficking of secreted factors and cell-surface receptors: New pathways to regulate cell proliferation and differentiation, and involvement in cancers. *Cell Commun Signal* **4**, 7 (2006).
47. Wesche, J., *et al.* FGF-1 and FGF-2 require the cytosolic chaperone Hsp90 for translocation into the cytosol and the cell nucleus. *J Biol Chem* **281**, 11405–11412 (2006).

48. Sorensen, V., *et al.* Different abilities of the four FGFRs to mediate FGF-1 translocation are linked to differences in the receptor C-terminal tail. *J Cell Sci* **119**, 4332–4341 (2006).
49. Bouleau, S., *et al.* FGF1 inhibits p53-dependent apoptosis and cell cycle arrest via an intracrine pathway. *Oncogene* **24**, 7839–7849 (2005).
50. Rodriguez-Enfedaque, A., *et al.* FGF1 nuclear translocation is required for both its neurotrophic activity and its p53-dependent apoptosis protection. *Biochim Biophys Acta* **1793**, 1719–1727 (2009).
51. Miller, D.L., Ortega, S., Bashayan, O., Basch, R., *et al.* Compensation by fibroblast growth factor 1 (FGF1) does not account for the mild phenotypic defects observed in FGF2 null mice [published erratum appears in Mol Cell Biol 2000 May;20(10):3752]. *Mol Cell Biol* **20**, 2260–2268 (2000).
52. Zhou, M., *et al.* Fibroblast growth factor 2 control of vascular tone. *Nat Med* **4**, 201–207 (1998).
53. Schultz, J.E., *et al.* Fibroblast growth factor-2 mediates pressure-induced hypertrophic response. *J Clin Invest* **104**, 709–719 (1999).
54. Amann, K., *et al.* Impaired myocardial capillarogenesis and increased adaptive capillary growth in FGF2-deficient mice. *Laboratory Investigation* **86**, 45–53 (2006).
55. Virag, J.A., *et al.* Fibroblast growth factor-2 regulates myocardial infarct repair: effects on cell proliferation, scar contraction, and ventricular function. *Am J Pathol* **171**, 1431–1440 (2007).
56. Nusayr, E. and Doetschman, T. Cardiac development and physiology are modulated by FGF2 in an isoform- and sex-specific manner. *Physiol Rep* **1**, 1–12 (2013).
57. Nusayr, E., Sadideen, D.T. and Doetschman, T. FGF2 modulates cardiac remodeling in an isoform- and sex-specific manner. *Physiol Rep* **1**, 1–14 (2013).
58. Guzy, R.D., Stoilov, I., Elton, T.J., Mecham, R.P., *et al.* Fibroblast growth factor 2 is required for epithelial recovery, but not for pulmonary fibrosis, in response to bleomycin. *Am J Respir Cell Mol Biol* **52**, 116–128 (2015).
59. Jonker, J.W., *et al.* A PPARgamma-FGF1 axis is required for adaptive adipose remodelling and metabolic homeostasis. *Nature* **485**, 391–394 (2012).
60. Suh, J.M., *et al.* Endocrinization of FGF1 produces a neomorphic and potent insulin sensitizer. *Nature* **513**, 436–439 (2014).

61. House, S.L., *et al.* Fibroblast Growth Factor 2 Mediates Isoproterenol-induced Cardiac Hypertrophy through Activation of the Extracellular Regulated Kinase. *Mol Cell Pharmacol* **2**, 143–154 (2010).
62. Dono, R., *et al.* FGF2 signaling is required for the development of neuronal circuits regulating blood pressure. *Circ Res* **90**, E5–E10 (2002).
63. Ortega, S., Ittmann, M., Tsang, S.H., Ehrlich, M., *et al.* Neuronal defects and delayed wound healing in mice lacking fibroblast growth factor 2. *Proc Natl Acad Sci USA* **95**, 5672–5677 (1998).
64. Grose, R. and Werner, S. Wound healing studies in transgenic and knockout mice. A review. *Methods Mol Med* **78**, 191–216 (2003).
65. Feldman, B., Poueymirou, W., Papaioannou, V.E., DeChiara, T.M., *et al.* Requirement of FGF-4 for postimplantation mouse development. *Science* **267**, 246–249 (1995).
66. Sun, X., *et al.* Conditional inactivation of FGF4 reveals complexity of signalling during limb bud development. *Nat Genet* **25**, 83–86 (2000).
67. Moon, A.M., Boulet, A.M. and Capecchi, M.R. Normal limb development in conditional mutants of FGF4. *Development* **127**, 989–996 (2000).
68. Sun, X., Mariani, F.V. and Martin, G.R. Functions of FGF signalling from the apical ectodermal ridge in limb development. *Nature* **418**, 501–508 (2002).
69. Naiche, L.A., Holder, N. and Lewandoski, M. FGF4 and FGF8 comprise the wavefront activity that controls somitogenesis. *Proc Natl Acad Sci USA* **108**, 4018–4023 (2011).
70. Hébert, J.M., Rosenquist, T., Götz, J. and Martin, G.R. FGF5 as a regulator of the hair growth cycle: Evidence from targeted and spontaneous mutations. *Cell* **78**, 1017–1025 (1994).
71. Cadieu, E., *et al.* Coat variation in the domestic dog is governed by variants in three genes. *Science* **326**, 150–153 (2009).
72. Drogemuller, C., Rufenacht, S., Wichert, B. and Leeb, T. Mutations within the FGF5 gene are associated with hair length in cats. *Anim Genet* **38**, 218–221 (2007).
73. Mulsant, P., Rochambeau, H. and De Thébault, R.G. A note on linkage between the angora and FGF5 genes in rabbits. *World Rabbit Sci.* **12** (2004).
74. Higgins, C.A., *et al.* FGF5 is a crucial regulator of hair length in humans. *Proc Natl Acad Sci USA* **111**, 10648–10653 (2014).

75. deLapeyriere, O., *et al.* Expression of the FGF6 gene is restricted to developing skeletal muscle in the mouse embryo. *Development* **118**, 601–611 (1993).
76. Floss, T., Arnold, H.H. and Braun, T. A role for FGF-6 in skeletal muscle regeneration. *Genes Dev* **11**, 2040–2051 (1997).
77. Fiore, F., Sebille, A. and Birnbaum, D. Skeletal muscle regeneration is not impaired in FGF6 −/− mutant mice. *Biochem Biophys Res Commun* **272**, 138–143 (2000).
78. Fiore, F., *et al.* Apparent normal phenotype of FGF6−/− mice. *Int J Devel Biol* **41**, 639–642 (1997).
79. Armand, A.S., *et al.* FGF6 regulates muscle differentiation through a calcineurin-dependent pathway in regenerating soleus of adult mice. *J Cell Physiol* **204**, 297–308 (2005).
80. Mansour, S.L., Goddard, J.M. and Capecchi, M.R. Mice homozygous for a targeted disruption of the proto-oncogene int-2 have developmental defects in the tail and inner ear. *Development* **117**, 13–28 (1993).
81. Alvarez, Y., *et al.* Requirements for FGF3 and FGF10 during inner ear formation. *Development* **130**, 6329–6338 (2003).
82. Tekin, M., *et al.* Homozygous mutations in fibroblast growth factor 3 are associated with a new form of syndromic deafness characterized by inner ear agenesis, microtia, and microdontia. *Am J Hum Genet* **80**, 338–344 (2007).
83. Tekin, M., *et al.* Homozygous FGF3 mutations result in congenital deafness with inner ear agenesis, microtia, and microdontia. *Clin Genet* **73**, 554–565 (2008).
84. Ramsebner, R., *et al.* A FGF3 mutation associated with differential inner ear malformation, microtia, and microdontia. *Laryngoscope* **120**, 359–364 (2010).
85. Alsmadi, O., *et al.* Syndromic congenital sensorineural deafness, microtia and microdontia resulting from a novel homoallelic mutation in fibroblast growth factor 3 (FGF3). *Eur J Hum Genet* **17**, 14–21 (2009).
86. Sensi, A., *et al.* LAMM syndrome with middle ear dysplasia associated with compound heterozygosity for FGF3 mutations. *Am J Med Genet A* **155a**, 1096–1197 (2011).
87. Terauchi, A., *et al.* Distinct FGFs promote differentiation of excitatory and inhibitory synapses. *Nature* **465**, 783–787 (2010).

88. Guo, L., Degenstein, L. and Fuchs, E. Keratinocyte growth factor is required for hair development but not for wound healing. *Genes Dev* **10**, 165–175 (1996).
89. Qiao, J., *et al.* FGF-7 modulates ureteric bud growth and nephron number in the developing kidney. *Development* **126**, 547–554 (1999).
90. Lee, C.H., Javed, D., Althaus, A.L., Parent, J.M., *et al.* Neurogenesis is enhanced and mossy fiber sprouting arises in FGF7-deficient mice during development. *Mol Cell Neurosci* **51**, 61–67 (2012).
91. Terauchi, A., *et al.* Selective synaptic targeting of the excitatory and inhibitory presynaptic organizers FGF22 and FGF7. *J Cell Sci* **128**, 281–292 (2015).
92. Takase, H.M., *et al.* FGF7 is a functional niche signal required for stimulation of adult liver progenitor cells that support liver regeneration. *Genes and Development* **27**, 169–181 (2013).
93. Brehm, J.M., *et al.* Identification of FGF7 as a novel susceptibility locus for chronic obstructive pulmonary disease. *Thorax* **66**, 1085–1090 (2011).
94. Min, H., *et al.* FGF-10 is required for both limb and lung development and exhibits striking functional similarity to *Drosophila* branchless. *Genes Dev* **12**, 3156–3161 (1998).
95. Sekine, K., *et al.* FGF10 is essential for limb and lung formation. *Nat Genet* **21**, 138–141 (1999).
96. Ohuchi, H., *et al.* FGF10 acts as a major ligand for FGF receptor 2 IIIb in mouse multi-organ development. *Biochem Biophys Res Commun* **277**, 643–649 (2000).
97. Sakaue, H., *et al.* Requirement of fibroblast growth factor 10 in development of white adipose tissue. *Genes Dev* **16**, 908–912 (2002).
98. Entesarian, M., *et al.* FGF10 missense mutations in aplasia of lacrimal and salivary glands (ALSG). *Eur J Hum Genet* **15**, 379–382 (2007).
99. Rohmann, E., *et al.* Mutations in different components of FGF signaling in LADD syndrome. *Nat Genet* **38**, 414–417 (2006).
100. Milunsky, J.M., Zhao, G., Maher, T.A., Colby, R., *et al.* LADD syndrome is caused by FGF10 mutations. *Clin Genet* **69**, 349–354 (2006).
101. Hsi, E., *et al.* A functional polymorphism at the FGF10 gene is associated with extreme myopia. *Invest Ophthalmol Vis Sci* **54**, 3265–3271 (2013).
102. Umemori, H., Linhoff, M.W., Ornitz, D.M. and Sanes, J.R. FGF22 and Its Close Relatives Are Presynaptic Organizing Molecules in the Mammalian Brain. *Cell* **118**, 257–570 (2004).

103. Lee, C.H. and Umemori, H. Suppression of epileptogenesis-associated changes in response to seizures in FGF22-deficient mice. *Front Cell Neurosci* **7**, 43 (2013).
104. Singh, R., *et al.* Fibroblast growth factor 22 contributes to the development of retinal nerve terminals in the dorsal lateral geniculate nucleus. *Front Mol Neurosci* **4**, 61 (2012).
105. Meyers, E.N., Lewandoski, M. and Martin, G.R. An FGF8 mutant allelic series generated by Cre- and Flp-mediated recombination. *Nat Genet* **18**, 136–141 (1998).
106. Sun, X., Meyers, E.N., Lewandoski, M. and Martin, G.R. Targeted disruption of FGF8 causes failure of cell migration in the gastrulating mouse embryo. *Genes Dev* **13**, 1834–1846 (1999).
107. Lewandoski, M., Sun, X. and Martin, G.R. FGF8 signalling from the AER is essential for normal limb development. *Nat Genet* **26**, 460–463 (2000).
108. Xu, J., Liu, Z. and Ornitz, D.M. Temporal and spatial gradients of FGF8 and FGF17 regulate proliferation and differentiation of midline cerebellar structures. *Development* **127**, 1833–1843 (2000).
109. Cholfin, J.A. and Rubenstein, J.L. Patterning of frontal cortex subdivisions by FGF17. *Proc Natl Acad Sci USA* **104**, 7652–7657 (2007).
110. Ohbayashi, N., *et al.* FGF18 is required for normal cell proliferation and differentiation during osteogenesis and chondrogenesis. *Genes & Development* **16**, 870–879 (2002).
111. Liu, Z., Lavine, K.J., Hung, I.H. and Ornitz, D.M. FGF18 is required for early chondrocyte proliferation, hypertrophy and vascular invasion of the growth plate. *Dev Biol* **302**, 80–91 (2007).
112. Liu, Z., Xu, J., Colvin, J.S. and Ornitz, D.M. Coordination of chondrogenesis and osteogenesis by fibroblast growth factor 18. *Genes Dev* **16**, 859–869 (2002).
113. Usui, H., *et al.* FGF18 is required for embryonic lung alveolar development. *Biochem Biophys Res Commun* **322**, 887–892 (2004).
114. Hasegawa, H., *et al.* Laminar patterning in the developing neocortex by temporally coordinated fibroblast growth factor signaling. *J Neurosci* **24**, 8711–8719 (2004).
115. Trarbach, E.B., *et al.* Nonsense mutations in FGF8 gene causing different degrees of human gonadotropin-releasing deficiency. *J Clin Endocrinol Metab* **95**, 3491–3496 (2010).
116. Miraoui, H., *et al.* Mutations in FGF17, IL17RD, DUSP6, SPRY4, and FLRT3 Are Identified in Individuals with Congenital Hypogonadotropic Hypogonadism. *Am J Hum Genet* **92**, 725–743 (2013).

117. Riley, B.M., *et al.* Impaired FGF signaling contributes to cleft lip and palate. *Proc Natl Acad Sci USA* **104**, 4512–4517 (2007).
118. Miyakawa, K., *et al.* A hydrophobic region locating at the center of fibroblast growth factor-9 is crucial for its secretion. *J Biol Chem* **274**, 29352–29357 (1999).
119. Revest, J.M., DeMoerlooze, L. and Dickson, C. Fibroblast growth factor 9 secretion is mediated by a non-cleaved amino-terminal signal sequence. *J Biol Chem* **275**, 8083–8090 (2000).
120. Miyakawa, K. and Imamura, T. Secretion of FGF-16 requires an uncleaved bipartite signal sequence. *J Biol Chem* **278**, 35718–35724 (2003).
121. Colvin, J.S., White, A., Pratt, S.J. and Ornitz, D.M. Lung hypoplasia and neonatal death in FGF9-null mice identify this gene as an essential regulator of lung mesenchyme. *Development* **128**, 2095–2106 (2001).
122. Pirvola, U., Zhang, X., Mantela, J., Ornitz, D.M., *et al.* FGF9 signaling regulates inner ear morphogenesis through epithelial-mesenchymal interactions. *Dev Biol* **273**, 350–360 (2004).
123. Colvin, J.S., Green, R.P., Schmahl, J., Capel, B. *et al.* Male-to-female sex reversal in mice lacking fibroblast growth factor 9. *Cell* **104**, 875–889 (2001b).
124. Dinapoli, L., Batchvarov, J. and Capel, B. FGF9 promotes survival of germ cells in the fetal testis. *Development* **133**, 1519–1527 (2006).
125. Hung, I.H., Yu, K., Lavine, K.J. and Ornitz, D.M. FGF9 regulates early hypertrophic chondrocyte differentiation and skeletal vascularization in the developing stylopod. *Dev Biol* **307**, 300–313 (2007).
126. Geske, M.J., Zhang, X., Patel, K.K., Ornitz, D.M. and Stappenbeck, T.S. FGF9 signaling regulates small intestinal elongation and mesenchymal development. *Development* **135**, 2959–2968 (2008).
127. Zhang, X., *et al.* Reciprocal epithelial-mesenchymal FGF signaling is required for cecal development. *Development* **133**, 173–180 (2006).
128. Lavine, K.J., *et al.* Endocardial and epicardial derived FGF signals regulate myocardial proliferation and differentiation *in vivo*. *Dev Cell* **8**, 85–95 (2005).
129. Hotta, Y., *et al.* FGF16 is required for cardiomyocyte proliferation in the mouse embryonic heart. *Dev Dyn* **237**, 2947–2954 (2008).
130. Lu, S.Y., *et al.* FGF-16 is required for embryonic heart development. *Biochem Biophys Res Commun* **373**, 270–274 (2008).
131. Matsumoto, E., *et al.* Angiotensin II-induced cardiac hypertrophy and fibrosis are promoted in mice lacking FGF16. *Genes Cells* **18**, 544–553 (2013).

132. Haara, O., *et al.* Ectodysplasin regulates activator-inhibitor balance in murine tooth development through FGF20 signaling. *Development* **139**, 3189–3199 (2012).
133. Huh, S.H., Jones, J., Warchol, M.E. and Ornitz, D.M. Differentiation of the lateral compartment of the cochlea requires a temporally restricted FGF20 signal. *PLoS Biol* **10**, e1001231 (2012).
134. Huh, S.H., *et al.* FGF20 governs formation of primary and secondary dermal condensations in developing hair follicles. *Genes Dev* **27**, 450–458 (2013).
135. Barak, H., *et al.* FGF9 and FGF20 maintain the stemness of nephron progenitors in mice and man. *Dev Cell* **22**, 1191–1207 (2012).
136. Huh, S.H., Warchol, M.E. and Ornitz, D.M. Cochlear progenitor number is controlled through mesenchymal FGF receptor signaling. *Elife* **4** (2015).
137. Wu, X.L., *et al.* Multiple synostoses syndrome is due to a missense mutation in exon 2 of FGF9 gene. *Am J Hum Genet* **85**, 53–63 (2009).
138. Harada, M., *et al.* FGF9 monomer/dimer equilibrium regulates extracellular matrix affinity and tissue diffusion. *Nature Genet* **41**, 289–298 (2009).
139. Chung, C.L., *et al.* Association of aberrant expression of sex-determining gene fibroblast growth factor 9 with sertoli cell-only syndrome. *Fertil Steril* **100**, 1547–54.e1–4 (2013).
140. Jones, B., Byers, H., Watson, J.S. and Newman, W.G. Identification of a novel familial FGF16 mutation in metacarpal 4-5 fusion. *Clin Dysmorphol* **23**, 95–97 (2014).
141. Jamsheer, A., *et al.* Whole exome sequencing identifies FGF16 nonsense mutations as the cause of X-linked recessive metacarpal 4/5 fusion. *J Med Genet* **50**, 579–584 (2013).
142. Laurell, T., *et al.* Identification of three novel FGF16 mutations in X-linked recessive fusion of the fourth and fifth metacarpals and possible correlation with heart disease. *Mol Genet Genomic Med* **2**, 402–411 (2014).
143. Jamsheer, A., *et al.* Further evidence for FGF16 truncating mutations as the cause of X-linked recessive fusion of metacarpals 4/5. *Birth Defects Res A Clin Mol Teratol* **100**, 314–318 (2014).
144. van der Walt, J.M., *et al.* Fibroblast growth factor 20 polymorphisms and haplotypes strongly influence risk of Parkinson disease. *Am J Hum Genet* **74**, 1121–1127 (2004).

145. Itoh, N. and Ohta, H. Roles of FGF20 in dopaminergic neurons and Parkinson's disease. *Front Mol Neurosci* **6**, 15 (2013).
146. Lemaitre, H., *et al.* Genetic variation in FGF20 modulates hippocampal biology. *J Neurosci* **30**, 5992–5997 (2010).
147. Haghnejad, L., *et al.* Variation in the miRNA-433 binding site of FGF20 is a risk factor for Parkinson's disease in Iranian population. *J Neurol Sci* **355**, 72–74 (2015).
148. Pan, J., *et al.* Fibroblast growth factor 20 (FGF20) polymorphism is a risk factor for Parkinson's disease in Chinese population. *Parkinsonism Relat Disord* **18**, 629–631 (2012).
149. de Mena, L., *et al.* FGF20 rs12720208 SNP and microRNA-433 variation: no association with Parkinson's disease in Spanish patients. *Neurosci Lett* **479**, 22–25 (2010).
150. Wider, C., *et al.* FGF20 and Parkinson's disease: No evidence of association or pathogenicity via alpha-synuclein expression. *Mov Disord* **24**, 455–459 (2009).
151. Wang, G., *et al.* Variation in the miRNA-433 binding site of FGF20 confers risk for Parkinson disease by overexpression of alpha-synuclein. *Am J Hum Genet* **82**, 283–289 (2008).
152. Clarimon, J., *et al.* Lack of evidence for a genetic association between FGF20 and Parkinson's disease in Finnish and Greek patients. *BMC Neurol* **5**, 11 (2005).
153. Gao, X., *et al.* Gene-gene interaction between FGF20 and MAOB in Parkinson disease. *Ann Hum Genet* **72**, 157–162 (2008).
154. Potthoff, M.J., Kliewer, S.A. and Mangelsdorf, D.J. Endocrine fibroblast growth factors 15/19 and 21: From feast to famine. *Genes Dev* **26**, 312–324 (2012).
155. Consortium, A. Autosomal dominant hypophosphataemic rickets is associated with mutations in FGF23. *Nat Genet* **26**, 345–348 (2000).
156. Goetz, R., *et al.* Molecular insights into the klotho-dependent, endocrine mode of action of fibroblast growth factor 19 subfamily members. *Mol Cell Biol* **27**, 3417–3428 (2007).
157. Angelin, B., Larsson, T.E. and Rudling, M. Circulating fibroblast growth factors as metabolic regulators — a critical appraisal. *Cell Metab* **16**, 693–705 (2012).
158. Goetz, R. and Mohammadi, M. Exploring mechanisms of FGF signalling through the lens of structural biology. *Nat Rev Mol Cell Biol* **14**, 166–80 (2013).

159. Kliewer, S.A. and Mangelsdorf, D.J. Fibroblast growth factor 21: from pharmacology to physiology. *Am J Clin Nutr* **91**, 254s–257s (2010).
160. Ding, X., *et al.* betaKlotho is required for fibroblast growth factor 21 effects on growth and metabolism. *Cell Metab* **16**, 387–393 (2012).
161. Suzuki, M., *et al.* betaKlotho is required for fibroblast growth factor (FGF) 21 signaling through FGF receptor (FGFR) 1c and FGFR3c. *Mol Endocrinol* **22**, 1006–1014 (2008).
162. Fon Tacer, K., *et al.* Research resource: Comprehensive expression atlas of the fibroblast growth factor system in adult mouse. *Mol Endocrinol* **24**, 2050–2064 (2010).
163. Wu, X., *et al.* FGF19-induced hepatocyte proliferation is mediated through FGFR4 activation. *J Biol Chem* **285**, 5165–5170 (2010).
164. Kurosu, H., *et al.* Regulation of fibroblast growth factor-23 signaling by klotho. *J Biol Chem* **281**, 6120–6123 (2006).
165. Nabeshima, Y. The discovery of alpha-Klotho and FGF23 unveiled new insight into calcium and phosphate homeostasis. *Cell Mol Life Sci* **65**, 3218–3230 (2008).
166. Gattineni, J., Twombley, K., Goetz, R., Mohammadi, M., *et al.* Regulation of serum 1,25(OH)2 vitamin D3 levels by fibroblast growth factor 23 is mediated by FGF receptors 3 and 4. *Am J Physiol Renal Physiol* **301**, F371–F377 (2011).
167. Gattineni, J., *et al.* Regulation of renal phosphate transport by FGF23 is mediated by FGFR1 and FGFR4. *Am J Physiol Renal Physiol* **306**, F351-F358 (2014).
168. Urakawa, I., *et al.* Klotho converts canonical FGF receptor into a specific receptor for FGF23. *Nature* **444**, 770–774 (2006).
169. Yu, C., *et al.* Elevated cholesterol metabolism and bile acid synthesis in mice lacking membrane tyrosine kinase receptor FGFR4. *J Biol Chem* **275**, 15482–15489 (2000).
170. Ho, H.K., *et al.* Fibroblast growth factor receptor 4 regulates proliferation, anti-apoptosis and alpha-fetoprotein secretion during hepatocellular carcinoma progression and represents a potential target for therapeutic intervention. *J Hepatol* **50**, 118–127 (2009).
171. Inagaki, T., *et al.* Fibroblast growth factor 15 functions as an enterohepatic signal to regulate bile acid homeostasis. *Cell Metab* **2**, 217–225 (2005).
172. Vincentz, J.W., McWhirter, J.R., Murre, C., Baldini, A., *et al.* FGF15 is required for proper morphogenesis of the mouse cardiac outflow tract. *Genesis* **41**, 192–201 (2005).

173. Wright, T.J., *et al.* Mouse FGF15 is the ortholog of human and chick FGF19, but is not uniquely required for otic induction. *Dev Biol* **269**, 264–275 (2004).
174. Uriarte, I., *et al.* Identification of fibroblast growth factor 15 as a novel mediator of liver regeneration and its application in the prevention of post-resection liver failure in mice. *Gut* **62**, 899–910 (2013).
175. Kong, B., *et al.* Fibroblast growth factor 15 deficiency impairs liver regeneration in mice. *Am J Physiol Gastrointest Liver Physiol* **306**, G893–G902 (2014).
176. French, D.M., *et al.* Targeting FGFR4 inhibits hepatocellular carcinoma in preclinical mouse models. *PLoS One* **7**, e36713 (2012).
177. Wu, A.L., *et al.* Amelioration of type 2 diabetes by antibody-mediated activation of fibroblast growth factor receptor 1. *Sci Transl Med* **3**, 113ra126 (2011).
178. Badman, M.K., *et al.* Hepatic fibroblast growth factor 21 is regulated by PPARalpha and is a key mediator of hepatic lipid metabolism in ketotic states. *Cell Metab* **5**, 426–437 (2007).
179. Inagaki, T., *et al.* Endocrine regulation of the fasting response by PPARalpha-mediated induction of fibroblast growth factor 21. *Cell Metab* **5**, 415–425 (2007).
180. Lundasen, T., *et al.* PPARalpha is a key regulator of hepatic FGF21. *Biochem Biophys Res Commun* **360**, 437–440 (2007).
181. Hotta, Y., *et al.* Fibroblast growth factor 21 regulates lipolysis in white adipose tissue but is not required for ketogenesis and triglyceride clearance in liver. *Endocrinology* **150**, 4625–4633 (2009).
182. Badman, M.K., Koester, A., Flier, J.S., Kharitonenkov, A., *et al.* Fibroblast growth factor 21-deficient mice demonstrate impaired adaptation to ketosis. *Endocrinology* **150**, 4931–4940 (2009).
183. Chu, A.Y., *et al.* Novel locus including FGF21 is associated with dietary macronutrient intake. *Hum Mol Genet* **22**, 1895–1902 (2013).
184. Tanaka, T., *et al.* Genome-wide meta-analysis of observational studies shows common genetic variants associated with macronutrient intake. *Am J Clin Nutr* **97**, 1395–1402 (2013).
185. Zhang, M., Zeng, L., Wang, Y.J., An, Z.M., *et al.* Associations of fibroblast growth factor 21 gene 3' untranslated region single-nucleotide polymorphisms with metabolic syndrome, obesity, and diabetes in a Han Chinese population. *DNA Cell Biol* **31**, 547–552 (2012).
186. Quarles, L.D. Skeletal secretion of FGF-23 regulates phosphate and vitamin D metabolism. *Nat Rev Endocrinol* **8**, 276–286 (2012).

187. Shimada, T., *et al.* Targeted ablation of FGF23 demonstrates an essential physiological role of FGF23 in phosphate and vitamin D metabolism. *J Clin Invest* **113**, 561–568 (2004).
188. Faul, C., *et al.* FGF23 induces left ventricular hypertrophy. *J Clin Invest* **121**, 4393–4408 (2011).
189. Touchberry, C.D., *et al.* FGF23 is a novel regulator of intracellular calcium and cardiac contractility in addition to cardiac hypertrophy. *Am J Physiol Endocrinol Metab* **304**, E863–E873 (2013).
190. Shimada, T., *et al.* Cloning and characterization of FGF23 as a causative factor of tumor-induced osteomalacia. *Proc Natl Acad Sci USA* **98**, 6500–6505 (2001).
191. Benet-Pages, A., Orlik, P., Strom, T.M. and Lorenz-Depiereux, B. An FGF23 missense mutation causes familial tumoral calcinosis with hyperphosphatemia. *Hum Mol Genet* **14**, 385–390 (2005).
192. Masi, L., *et al.* A novel recessive mutation of fibroblast growth factor-23 in tumoral calcinosis. *J Bone Joint Surg Am* **91**, 1190–1198 (2009).
193. Masi, L., *et al.* Can fibroblast growth factor (FGF)-23 circulating levels suggest coronary artery abnormalities in children with Kawasaki disease? *Clin Exp Rheumatol* **31**, 149–153 (2013).
194. Falcini, F., *et al.* Fibroblast growth factor 23 (FGF23) gene polymorphism in children with Kawasaki syndrome (KS) and susceptibility to cardiac abnormalities. *Ital J Pediatr* **39**, 69 (2013).
195. Smallwood, P.M., *et al.* Fibroblast growth factor (FGF) homologous factors: new members of the FGF family implicated in nervous system development. *Proc Natl Acad Sci USA* **93**, 9850–9857 (1996).
196. Pablo, J.L. and Pitt, G.S. Fibroblast Growth Factor Homologous Factors: New Roles in Neuronal Health and Disease. *Neuroscientist* **22**, 19–25 (2016).
197. Hsu, W.C., Nilsson, C.L. and Laezza, F. Role of the axonal initial segment in psychiatric disorders: function, dysfunction, and intervention. *Front Psychiatry* **5**, 109 (2014).
198. Xiao, M., Bosch, M.K., Nerbonne, J.M. and Ornitz, D.M. FGF14 localization and organization of the axon initial segment. *Mol Cell Neurosci* **56**, 393–403 (2013).
199. Goldfarb, M., *et al.* Fibroblast growth factor homologous factors control neuronal excitability through modulation of voltage-gated sodium channels. *Neuron* **55**, 449–463 (2007).

200. Lou, J.Y., *et al.* Fibroblast growth factor 14 is an intracellular modulator of voltage-gated sodium channels. *J Physiol* **569**, 179–193 (2005).
201. Wang, C., *et al.* Fibroblast growth factor homologous factor 13 regulates Na+ channels and conduction velocity in murine hearts. *Circ Res* **109**, 775–782 (2011).
202. Wildburger, N.C., *et al.* Quantitative proteomics reveals protein-protein interactions with fibroblast growth factor 12 as a component of the voltage-gated sodium channel 1.2 (nav1.2) macromolecular complex in Mammalian brain. *Mol Cell Proteomics* **14**, 1288–1300 (2015).
203. Schoorlemmer, J. and Goldfarb, M. Fibroblast growth factor homologous factors are intracellular signaling proteins. *Curr Biol* **11**, 793–797. (2001).
204. Wu, Q.F., *et al.* Fibroblast growth factor 13 is a microtubule-stabilizing protein regulating neuronal polarization and migration. *Cell* **149**, 1549–1564 (2012).
205. Wang, Q., *et al.* Ataxia and paroxysmal dyskinesia in mice lacking axonally transported FGF14. *Neuron* **35**, 25–38 (2002).
206. Wozniak, D.F., Xiao, M., Xu, L., Yamada, K.A., *et al.* Impaired spatial learning and defective theta burst induced LTP in mice lacking fibroblast growth factor 14. *Neurobiol Dis* **26**, 14–26 (2007).
207. Xiao, M., *et al.* Impaired hippocampal synaptic transmission and plasticity in mice lacking fibroblast growth factor 14. *Mol Cell Neurosci* **34**, 366–377 (2007).
208. Liu, C., Dib-Hajj, S.D. and Waxman, S.G. Fibroblast growth factor homologous factor 1B binds to the C terminus of the tetrodotoxin-resistant sodium channel rNav1.9a (NaN). *J Biol Chem* **276**, 18925–18933. (2001).
209. Liu, C.J., Dib-Hajj, S.D., Renganathan, M., Cummins, T.R., *et al.* Modulation of the cardiac sodium channel Na(v)1.5 by fibroblast growth factor homologous factor 1B. *J Biol Chem* **278**, 1029–1036 (2003).
210. Laezza, F., *et al.* FGF14 n-terminal splice variants differentially modulate nav1.2 and nav1.6-encoded sodium channels. *Mol Cell Neurosci* **42**, 90–101 (2009).
211. Shavkunov, A., *et al.* Bioluminescence methodology for the detection of protein-protein interactions within the voltage-gated sodium channel macromolecular complex. *Assay Drug Dev Technol* **10**, 148–160 (2012).

212. Hennessey, J.A., Wei, E.Q. and Pitt, G.S. Fibroblast growth factor homologous factors modulate cardiac calcium channels. *Circ Res* **113**, 381–388 (2013).
213. Dover, K., Solinas, S., D'Angelo, E. and Goldfarb, M. Long-Term Inactivation Particle for Voltage-Gated Sodium Channels. *J Physiol* **588**, 3695–3711 (2010).
214. Shakkottai, V.G., *et al.* FGF14 regulates the intrinsic excitability of cerebellar Purkinje neurons. *Neurobiol Dis* **33**, 81–88 (2009).
215. Bosch, M.K., *et al.* Intracellular FGF14 (iFGF14) is required for spontaneous and evoked firing in cerebellar purkinje neurons and for motor coordination and balance. *J Neurosci* **35**, 6752–6769 (2015).
216. Shavkunov, A.S., *et al.* The fibroblast growth factor 14.voltage-gated sodium channel complex is a new target of glycogen synthase kinase 3 (GSK3). *J Biol Chem* **288**, 19370–19385 (2013).
217. Wildburger, N.C. and Laezza, F. Control of neuronal ion channel function by glycogen synthase kinase-3: New prospective for an old kinase. *Front Mol Neurosci* **5**, 80 (2012).
218. Ali, S., Shavkunov, A., Panova, N., Stoilova-McPhie, S., *et al.* Modulation of the FGF14:FGF14 Homodimer Interaction through Short Peptide Fragments. *CNS Neurol Disord Drug Targets* **13**, 1559–1570 (2014).
219. Hsu, W.C., *et al.* Identifying a Kinase Network Regulating FGF14:Nav1.6 Complex Assembly Using Split-Luciferase Complementation. *PMC Biophys* **10**, e0117246 (2015).
220. Hennessey, J.A., *et al.* FGF12 is a candidate Brugada syndrome locus. *Heart Rhythm* **10**, 1886–1894 (2013).
221. Gecz, J., *et al.* Fibroblast growth factor homologous factor 2 (FHF2): Gene structure, expression and mapping to the Borjeson-Forssman-Lehmann syndrome region in Xq26 delineated by a duplication breakpoint in a BFLS-like patient. *Hum Genet* **104**, 56–63 (1999).
222. DeStefano, G.M., *et al.* Position effect on FGF13 associated with X-linked congenital generalized hypertrichosis. *Proc Natl Acad Sci USA* **110**, 7790–7795 (2013).
223. Van Swieten, J.C., *et al.* A mutation in the fibroblast growth factor 14 gene is associated with autosomal dominant cerebellar ataxia. *Am J Hum Genet* **72**, 191–199 (2003).
224. Brusse, E., *et al.* Spinocerebellar ataxia associated with a mutation in the fibroblast growth factor 14 gene (SCA27): A new phenotype. *Mov Disord* **21**, 396–401 (2006).

225. Choquet, K., La Piana, R. and Brais, B. A novel frameshift mutation in FGF14 causes an autosomal dominant episodic ataxia. *Neurogenetics* **16**, 233–236 (2015).
226. Misceo, D., *et al.* SCA27 caused by a chromosome translocation: Further delineation of the phenotype. *Neurogenetics* **10**, 371–374 (2009).
227. Shimojima, K., *et al.* Spinocerebellar ataxias type 27 derived from a disruption of the fibroblast growth factor 14 gene with mimicking phenotype of paroxysmal non-kinesigenic dyskinesia. *Brain Dev* **34**, 230–233 (2012).
228. Laezza, F., *et al.* The FGF14(F145S) mutation disrupts the interaction of FGF14 with voltage-gated Na+ channels and impairs neuronal excitability. *J Neurosci* **27**, 12033–12044 (2007).
229. Raballo, R., *et al.* Basic fibroblast growth factor (FGF2) is necessary for cell proliferation and neurogenesis in the developing cerebral cortex. *J Neurosci* **20**, 5012–5023 (2000).
230. Montero, A., *et al.* Disruption of the fibroblast growth factor-2 gene results in decreased bone mass and bone formation. *J Clin Invest* **105**, 1085–1093 (2000).
231. Timmer, M., *et al.* Fibroblast growth factor (FGF)-2 and FGF receptor 3 are required for the development of the substantia nigra, and FGF-2 plays a crucial role for the rescue of dopaminergic neurons after 6-hydroxydopamine lesion. *J Neurosci* **27**, 459–471 (2007).
232. Homer-Bouthiette, C., Doetschman, T., Xiao, L. and Hurley, M.M. Knockout of nuclear high molecular weight FGF2 isoforms in mice modulates bone and phosphate homeostasis. *J Biol Chem* **289**, 36303–36314 (2014).
233. Urness, L.D., Bleyl, S.B., Wright, T.J., Moon, A.M., *et al.* Redundant and dosage sensitive requirements for FGF3 and FGF10 in cardiovascular development. *Dev Biol* **356**, 383–397 (2011).
234. Parker, H.G., *et al.* An expressed FGF4 retrogene is associated with breed-defining chondrodysplasia in domestic dogs. *Science* **325**, 995–998 (2009).
235. Neuhaus, P., *et al.* Reduced mobility of fibroblast growth factor (FGF)-deficient myoblasts might contribute to dystrophic changes in the musculature of FGF2/FGF6/mdx triple-mutant mice. *Mol Cell Biol* **23**, 6037–6048 (2003).
236. Alpdogan, O., *et al.* Keratinocyte growth factor (KGF) is required for postnatal thymic regeneration. *Blood* **107**, 2453–2460 (2006).

237. Peng, C., He, Q. and Luo, C. Lack of keratinocyte growth factor retards angiogenesis in cutaneous wounds. *J Int Med Res* **39**, 416–423 (2011).
238. Moon, A.M. and Capecchi, M.R. FGF8 is required for outgrowth and patterning of the limbs. *Nat Genet* **26**, 455–459 (2000).
239. Brown, C.B., *et al.* Cre-mediated excision of FGF8 in the Tbx1 expression domain reveals a critical role for FGF8 in cardiovascular development in the mouse. *Dev Biol* **267**, 190–192 (2004).
240. Perantoni, A.O., *et al.* Inactivation of FGF8 in early mesoderm reveals an essential role in kidney development. *Development* **132**, 3859–3871 (2005).
241. Ilagan, R., *et al.* FGF8 is required for anterior heart field development. *Development* **133**, 2435–2445 (2006).
242. Park, E.J., *et al.* Required, tissue-specific roles for FGF8 in outflow tract formation and remodeling. *Development* **133**, 2419–2433 (2006).
243. Jacques, B.E., Montcouquiol, M.E., Layman, E.M., Lewandoski, M., *et al.* FGF8 induces pillar cell fate and regulates cellular patterning in the mammalian cochlea. *Development* **134**, 3021–3029 (2007).
244. Zelarayan, L.C., *et al.* Differential requirements for FGF3, FGF8 and FGF10 during inner ear development. *Dev Biol* **308**, 379–391 (2007).
245. Ladher, R.K., Wright, T.J., Moon, A.M., Mansour, S.L., *et al.* FGF8 initiates inner ear induction in chick and mouse. *Genes Dev* **19**, 603–613 (2005).
246. Hasegawa, K. and Saga, Y. FGF8-FGFR1 Signaling Acts as a Niche Factor for Maintaining Undifferentiated Spermatogonia in the Mouse. *Biol Reprod* **91**, 145 (2014).
247. Schmahl, J., Kim, Y., Colvin, J.S., Ornitz, D.M., *et al.* FGF9 induces proliferation and nuclear localization of FGFR2 in Sertoli precursors during male sex determination. *Development* **131**, 3627–3636 (2004).
248. White, A.C., *et al.* FGF9 and SHH signaling coordinate lung growth and development through regulation of distinct mesenchymal domains. *Development* **133**, 1507–1517 (2006).
249. Lin, Y., *et al.* Neuron-derived FGF9 is essential for scaffold formation of Bergmann radial fibers and migration of granule neurons in the cerebellum. *Dev Biol* **329**, 44–54 (2009).
250. Yin, Y., Wang, F. and Ornitz, D.M. Mesothelial- and epithelial-derived FGF9 have distinct functions in the regulation of lung development. *Development* **138**, 3169–3177 (2011).

251. Marguerie, A., *et al.* Congenital heart defects in FGFr2-IIIb and FGF10 mutant mice. *Cardiovasc Res* **71**, 50–60 (2006).
252. Nyeng, P., Norgaard, G.A., Kobberup, S. and Jensen, J. FGF10 signaling controls stomach morphogenesis. *Dev Biol* **303**, 295–310 (2006).
253. Abler, L.L., Mansour, S.L. and Sun, X. Conditional gene inactivation reveals roles for FGF10 and FGFr2 in establishing a normal pattern of epithelial branching in the mouse lung. *Dev Dynam* **238**, 1999–2013 (2009).
254. Bhushan, A., *et al.* FGF10 is essential for maintaining the proliferative capacity of epithelial progenitor cells during early pancreatic organogenesis. *Development* **128**, 5109–5117. (2001).
255. Wright, T.J. and Mansour, S.L. FGF3 and FGF10 are required for mouse otic placode induction. *Development* **130**, 3379–3390 (2003).
256. Rice, R., *et al.* Disruption of FGF10/FGFr2b-coordinated epithelial-mesenchymal interactions causes cleft palate. *J Clin Invest* **113**, 1692–1700 (2004).
257. Jaskoll, T., *et al.* FGF10/FGFR2b signaling plays essential roles during in vivo embryonic submandibular salivary gland morphogenesis. *BMC Dev Biol* **5**, 11 (2005).
258. Veltmaat, J.M., *et al.* Gli3-mediated somitic FGF10 expression gradients are required for the induction and patterning of mammary epithelium along the embryonic axes. *Development* **133**, 2325–2335 (2006).
259. Michos, O., *et al.* Kidney development in the absence of Gdnf and Spry1 requires FGF10. *PLoS Genet* **6**, e1000809 (2010).
260. Qu, X., *et al.* Lacrimal gland development and FGF10-FGFr2b signaling are controlled by 2-O- and 6-O-sulfated heparan sulfate. *J Biol Chem* **286**, 14435–14444 (2011).
261. Sala, F.G., *et al.* FGF10 controls the patterning of the tracheal cartilage rings via Shh. *Development* **138**, 273–282 (2011).
262. Sohn, W.J., *et al.* Reciprocal interactions of FGF10/FGFr2b modulate the mouse tongue epithelial differentiation. *Cell Tissue Res* **345**, 265–273 (2011).
263. Al Alam, D., *et al.* FGF9-Pitx2-FGF10 signaling controls cecal formation in mice. *Dev Biol* **369**, 340–348 (2012).
264. Urness, L.D., Wang, X., Shibata, S., Ohyama, T., *et al.* FGF10 is required for specification of non-sensory regions of the cochlear epithelium. *Dev Biol* **400**, 59–71 (2015).

265. Rochais, F., *et al.* FGF10 promotes regional foetal cardiomyocyte proliferation and adult cardiomyocyte cell-cycle re-entry. *Cardiovasc Res* **104**, 432–442 (2014).
266. Puranam, R.S., *et al.* Disruption of FGF13 causes synaptic excitatory-inhibitory imbalance and genetic epilepsy and febrile seizures plus. *J Neurosci* **35**, 8866–8881 (2015).
267. Choi, M., *et al.* Identification of a hormonal basis for gallbladder filling. *Nat Med* **12**, 1253–1255 (2006).
268. Borello, U., *et al.* FGF15 promotes neurogenesis and opposes FGF8 function during neocortical development. *Neural Dev* **3**, 17 (2008).
269. Uriarte, I., *et al.* Ileal FGF15 contributes to fibrosis-associated hepatocellular carcinoma development. *Int J Cancer* (2014).
270. Whitsett, J.A., *et al.* Fibroblast growth factor 18 influences proximal programming during lung morphogenesis. *J Biol Chem* **277**, 22743–22749 (2002).
271. Murata, Y., *et al.* FGF21 impairs adipocyte insulin sensitivity in mice fed a low-carbohydrate, high-fat ketogenic diet. *PLoS One* **8**, e69330 (2013).
272. Planavila, A., *et al.* Fibroblast growth factor 21 protects the heart from oxidative stress. *Cardiovasc Res* **106**, 19–31 (2014).
273. Jarosz, M., *et al.* Fibroblast growth factor 22 is not essential for skin development and repair but plays a role in tumorigenesis. *PLoS One* **7**, e39436 (2012).
274. Pasaoglu, T. and Schikorski, T. Presynaptic size of associational/commissural CA3 synapses is controlled by fibroblast growth factor 22 in adult mice. *Hippocampus* **26**, 151–160 (2015).
275. Hu, M.C., Shiizaki, K., Kuro-o, M. and Moe, O.W. Fibroblast growth factor 23 and Klotho: physiology and pathophysiology of an endocrine network of mineral metabolism. *Annu Rev Physiol* **75**, 503–533 (2013).
276. Sitara, D., *et al.* Homozygous ablation of fibroblast growth factor-23 results in hyperphosphatemia and impaired skeletogenesis, and reverses hypophosphatemia in Phex-deficient mice. *Matrix Biol* **23**, 421–432 (2004).
277. Lysaght, A.C., *et al.* FGF23 deficiency leads to mixed hearing loss and middle ear malformation in mice. *PLoS One* **9**, e107681 (2014).
278. Yuan, Q., *et al.* Increased osteopontin contributes to inhibition of bone mineralization in FGF23-deficient mice. *J Bone Miner Res* **29**, 693–704 (2014).
279. Gregory-Evans, C.Y., *et al.* SNP genome scanning localizes oto-dental syndrome to chromosome 11q13 and microdeletions at this locus

implicate FGF3 in dental and inner-ear disease and FADD in ocular coloboma. *Hum Mol Genet* **16**, 2482–2493 (2007).
280. Kehler, J.S., *et al.* Four independent mutations in the feline fibroblast growth factor 5 gene determine the long-haired phenotype in domestic cats. *J Hered* **98**, 555–566 (2007).
281. Dierks, C., Momke, S., Philipp, U. and Distl, O. Allelic heterogeneity of FGF5 mutations causes the long-hair phenotype in dogs. *Anim Genet* **44**, 425–431 (2013).
282. Falardeau, J., *et al.* Decreased FGF8 signaling causes deficiency of gonadotropin-releasing hormone in humans and mice. *J Clin Invest* **118**, 2822–2831 (2008).
283. Arauz, R.F., *et al.* A hypomorphic allele in the fgf8 gene contributes to holoprosencephaly and is allelic to gonadotropin-releasing hormone deficiency in humans. *Mol Syndromol* **1**, 59–66 (2010).
284. McCabe, M.J., *et al.* Novel FGF8 mutations associated with recessive holoprosencephaly, craniofacial defects, and hypothalamo-pituitary dysfunction. *J Clin Endocrinol Metab* **96**, E1709–E1718 (2011).
285. Valdes-Socin, H., *et al.* Reproduction, smell, and neurodevelopmental disorders: genetic defects in different hypogonadotropic hypogonadal syndromes. *Front Endocrinol (Lausanne)* **5**, 109 (2014).
286. Entesarian, M., *et al.* Mutations in the gene encoding fibroblast growth factor 10 are associated with aplasia of lacrimal and salivary glands. *Nat Genet* **37**, 125–127 (2005).
287. Zhang, F., *et al.* Exome sequencing identified FGF12 as a novel candidate gene for Kashin-Beck disease. *Funct Integr Genomics* **16**, 13–17 (2016).
288. Dalski, A., *et al.* Mutation analysis in the fibroblast growth factor 14 gene: frameshift mutation and polymorphisms in patients with inherited ataxias. *Eur J Hum Genet* **13**, 118–120 (2005).
289. Coebergh, J.A., *et al.* A new variable phenotype in spinocerebellar ataxia 27 (SCA 27) caused by a deletion in the FGF14 gene. *Eur J Paediatr Neurol* **18**, 413–415 (2014).
290. Chefetz, I., *et al.* A novel homozygous missense mutation in FGF23 causes familial tumoral calcinosis associated with disseminated visceral calcification. *Hum Genet* **118**, 261–266 (2005).
291. Araya, K., *et al.* A Novel Mutation in fibroblast growth factor (FGF)23 gene as a cause of tumoral calcinosis. *J Clin Endocrinol Metab* **90**, 5523–5527 (2005).
292. Garringer, H.J., *et al.* Molecular genetic and biochemical analyses of FGF23 mutations in familial tumoral calcinosis. *Am J Physiol Endocrinol Metab* **295**, E929–E937 (2008).

Chapter 2
Regulation of FGF Signaling

Robert E. Friesel*

Abstract

The fibroblast growth factors (FGFs) are a family of signaling proteins that play roles in development, physiology and disease. The FGFs activate a variety of signaling pathways through binding to one of four FGF receptors. These receptors are tyrosine kinases that activate several downstream signaling pathways. Research has show that FGF signaling pathways are subject to complex regulation by both inhibitors and activators of these signaling circuits. This chapter will describe current knowledge of the complex interplay of positive and negative regulators of FGF signaling.

1. Introduction

The fibroblast growth factors (FGF) are a family of signaling molecules that act either locally in a paracrine manner or as secreted factors that function as hormones (**see chapter** 1).[1–3] The FGFs elicit their effects through four tyrosine kinase receptors. FGFs have a wide range of context dependent effects on cells including proliferation, migration, differentiation and survival. FGFs play critical roles in development, tissue repair and homeostasis. Due to the

*Center for Molecular Medicine, Marine Medical Center Research Institute, Scarborough, ME 04074, USA; E-mail: friesr@mmc.org

number of ligands and receptors and receptor spice variants, FGF signaling is quite complex. The importance of tight regulation of FGF signaling is illustrated by the fact that dysregulation of FGF signaling results in several developmental disorders of the skeleton and reproductive system as well as being a contributor to the development of cancer and cardiovascular disease.[1,2,4–8]

FGF binding to its receptor results receptor dimerization and transphosphorylation on seven tyrosine residues.[1,9] These phosphotyrosine residues serve as binding sites for adaptor proteins such as Shc and Grb2 and other signaling molecules activating downstream signaling pathways such as extracellular signal-regulated kinase (ERK), phosphoinositide 3-kinase (PI3K), c-Src, and phospholipase Cγ (PLCγ). These pathways play roles in FGF induced cell proliferation, migration, differentiation and survival. The duration and intensity of these signals dictates in part the cellular response to FGF signaling. The duration and intensity of FGF signaling is regulated in part by feedback control mechanisms that act on various parts of the FGF signaling cascade. Recent research has revealed several mechanisms by which FGF signaling is either positively or negatively regulated thus determining the intensity and duration of the signaling output.[10,11] This chapter will cover recent studies that provide new insights in the complexity of FGF signaling regulation in development, tissue repair, and homeostasis.

2. Regulation of FGF Signaling by Sprouty

Sprouty (Spry) was originally identified as an inhibitor of FGF mediated tracheal branching during *Drosophila* development.[12] Four mammalian Spry genes were identified based upon sequence similarity to *Drosophila* Spry.[13] Mammalian Spry proteins are shorter that their *Drosophila* homolog, however they share a highly conserved C-terminal cysteine-rich domain (CRD) also called the cysteine-rich Sprouty-related (SPR) domain. The N-termini of the four mammalian Spry proteins have limited sequence homology suggesting a divergence in function.[14,15] The Sprys modulate receptor tyrosine kinase (RTK) signaling principally by inhibiting ERK signaling, however other pathways are affected as well.

The expression of Spry family members co-localizes with the expression of FGFs during mammalian development suggesting that FGFs may regulate their expression.[13,15] In adult tissues, the expression of Spry1, 2, and 4 is widespread, whereas Spry3 expression is restricted to brain and testes.[14,15] *In vivo* and *ex vivo* studies indicate that expression of Sprys is induced by the RTK pathways that they modulate, mainly FGF but also platelet-derived growth factor (PDGF),[16] vascular endothelial growth factor (VEGF),[17] epidermal growth factor (EGF)[18,19] and glial cell-derived neurotrophic factor (GDNF),[20,21] often in a cell context specific manner. The Spry promoters have not been extensively studied, however, functional FoxO, Ets-1, AP2, SP1, CREB and GATA binding sites as have been identified in the Spry2 promoter.[22] A preliminary characterization of the Spry4 promoter indicates that it contains binding sites for the transcription factors HIF1, STAT5, SP1, WT1, SREBP and PBX-1.[23] These initial promoter analyses indicate that Spry promoters have binding sites for growth factor and hypoxia responsive transcription factors. Definitive studies on *in vivo* transcriptional regulation of Spry expression require further studies.

Spry proteins modulate RTK signaling, the best studied of which are the FGF receptors (FGFR) and EGF receptor (EGFR). *In vitro* studies of Spry function have focused on their roles in RTK mediated cell proliferation and migration. Several studies have shown that forced expression of Spry1, Spry2 and Spry4 inhibit FGF induced ERK activation, cell proliferation and migration in several cell types.[14,15] The extent to which FGF signaling is inhibited depends on the cell type and the Spry family member that is over expressed. Forced expression of Spry4 in endothelial cells resulted in cell cycle arrest in the G_1/G_2 phase of the cell cycle accompanied by decreased phosphorylation of the cell cycle regulatory protein Rb, as well as an increase in the cell. Cycle inhibitors p21 and p27.[24] Spry4 also inhibits FGF induced angiogenesis *in vitro* and *in vivo*.[24] Recent studies have shown that Spry1 and Spry2 inhibit FGF induced tyrosine phosphorylation of PLCγ and decreased production of inositol (1,4,5)-triphosphate (IP_3) and diacylglycerol as well as inhibited calcium mediated signaling downstream of PLCγ.[25] Furthermore,

Spry2 inhibits FGF mediated activation of a PKCδ-PKD1 complex, which is another means by which ERK can be activated.[26]

Several studies have shown that Spry2 has several sites of action in the Ras/ERK pathway including upstream of Ras by interfering with Grb2-Sos complex activation, or at the level of Raf1.[14,15] To add to the complexity of Ras/ERK modulation by Spry, several Spry binding partners have been identified including Grb2,[27,28] c-Cbl,[18,29,30] Tesk1,[31–33] PP2A,[34] CIN85,[35] caveolin1,[36] DYRK1A,[37] Raf1[17,38] and SIAH2.[39,40] Spry2 contains a tyrosine residue that is highly conserved among all Spry family members (Tyr55 in Spry2) that becomes phosphorylated upon RTK activation by Src family kinases.[41] Phosphorylation of this residue results in a conformational change that reveals a cryptic SH3 binding motif (PXXPXR) at the C-terminus of Spry2, which serves as a binding site for Grb2.[28] By binding Grb2, Spry2 sequesters it away from Sos thus limiting the activation of the ERK pathway. This SH3 binding motif is absent in other Spry family members and suggests a possible explanation as to why Spry2 is more potent at inhibiting FGFR signaling than Spry1.[28] Mutation of Tyr55 on Spry2 abrogates its inhibitory action on FGFR mediated ERK signaling. Phosphorylation of Tyr55 of Spry2 is essential for the interaction of Spry2 with the E3 ubiquitin ligase c-Cbl through its SH2 domain.[29] Binding of c-Cbl to Spry2 results in proteasomal degradation of Spry2 thus limiting its inhibitory action on FGFR-ERK signaling.[30] The tyrosine phosphatase SHP2 dephosphorylates Spry2 at Tyr55, and may explain in part the observation that overexpression of SHP2 enhances FGFR signaling.[42] A single report also suggests that phosphorylation of Tyr227 of Spry2 is also require to inhibit FGFR signaling.[43]

Serine and threonine phosphorylation are also important to regulating Spry function in the context of FGF signaling. FGFR activation results in the dephosphorylation of Ser112 and Ser115 by PP2A and this dephosphorylation may be a necessary step for tyrosine phosphorylation of Tyr55.[34] The serine/threonine kinase Mnk1 phosphorylates Spry2 on Ser112 and Ser121, and mutation of these residues to Ala results in increased phosphorylation of Tyr55 and accelerated degradation of Spry2.[44] The dual-specificity tyrosine-phosphorylated and -regulated kinase 1A (DYRK1A) interacts with and phosphorylates Spry2 on Thr75.[37] This phosphorylation seems to negatively impact the function of Spry2 as negative regulator of FGF signaling. In addition, testicular protein kinase 1(TESK1), which phosphorylates coflin and regulates actin cytoskeleton dynamics, has been shown to bind Spry4.[32] This interaction occurs through the C-terminal domain of Spry4 and inhibits the kinase activity of TESK1 towards cofilin.[33] In addition, TESK1 inhibits the action of Spry2 on the ERK pathway independent of its kinase activity, by sequestering Spry2 away from Grb2.[31]

Stimulation of cells with FGF results in Spry2 translocation to the plasma membrane, which is necessary for its inhibitory function on proliferation and migration.[14,15] Studies have determined that the C-terminal cysteine-rich domain is necessary for this translocation, and several mechanisms for this localization have been proposed. One, the C-terminal cysteine-rich domain of Spry2 binds to phosphatidylinositol 4,5-bisphosphate, a constituent of the plasma membrane.[45] In endothelial cells, palmitoylation of cysteine-rich domain anchors Spry1 and Spry2 in the membrane, although whether this is regulated by growth factor stimulation requires further study.[36] Spry proteins also translocate to the membrane and associate with caveolin-1 through the cysteine-rich domain after FGF-stimulated serine phosphorylation of caveolin-1.[36,46] Different Spry family members vary greatly in their cooperation with caveolin-1 to inhibit ERK activation by FGF2 and other growth factors.

Several reports indicate that Sprys can form homo- and hetero-oligomers.[39,47] All four Spry isoforms interact with one another through their cysteine-rich domains. Different Spry heterodimers bind different combinations of Spry interacting proteins.[47] A heterodimer of Spry1 and Spry4 was the most potent heterodimer for inhibiting FGF-induced ERK activation, where Spry1 binds Grb2 and Spry4 binds Sos.[47] The ratio and abundance of various Spry oligomers likely influences the degree to which FGF signaling is inhibited and is likely regulated at both the transcriptional and post-translational levels.

Most structural and functional studies have been performed on Spry2, and much remains to be learned about the functions of Spry1 and Spry4. Gene targeting studies in the mouse suggests that Sprys have both redundant and non-redundant functions in a context dependent manner.[20,48–52] In support of this notion, it was recently reported that Spry1 and Spry4 had opposing functions in regulating vascular smooth muscle cell phenotype.[53] Further investigation is required to determine how Spry isoforms regulate the diversity of signaling outputs of FGFR signaling.

3. Regulation of FGF Signaling by Sef

Sef (similar expression to FGF) was identified in zebrafish embryos as an inhibitor of FGF signaling and ERK activation.[54,55] Similar to Sprys, Sef is expressed in areas of FGF signaling during vertebrate embryogenesis, and FGF stimulation induces the expression of Sef in a variety of cells and embryonic tissues and thus is a feedback inhibitor of FGF signaling. Sef was identified in humans as a novel cell surface protein expressed in human umbilical vein endothelial cells (HUVEC), and in highly vascularized tissues such as kidney, heart, skeletal muscle and colon.[56] Sef was also expressed in ductal epithelial cells of several organs, and in breast cancer tissues. Sef is a type I transmembrane protein that has an intracellular domain with a high degree of sequence similarity to IL-17 receptors and this intracellular domain has been named the SEF/IL-17R (SEFIR) domain.

Sef has been shown to inhibit FGF signaling in zebrafish and *Xenopus* embryos,[54,55] and in several cell culture models including HUVEC,[57] PC12,[58] and NIH3T3 cells.[59] Sef was shown to physically associate with FGFR1 and FGFR2 by co-immunoprecipitation and this association resulted in the inhibition of both ERK and Akt activation.[54,59] The interaction is mediated in part by the transmembrane domain of Sef, because replacement of the transmembrane domain of Sef with the PDGF receptor transmembrane domain abrogated the interaction with FGFR1 and the inhibition of ERK activation.[57] In humans, alternative splicing of Sef transcripts gives rise to an isoform of Sef called hSef-b that contains all but the first 42 residues found in hSef-a, which are replaced with 10 different amino acids.[60] The hSef-b isoform lacks a signal sequence for membrane insertion and presumably is retained in the cytoplasm or ER-Golgi. The hSef-b encoded protein inhibits FGF induced ERK activation without affecting Akt activation. The hSef-b isoform also inhibited PDGF induced ERK activation, but not other RTKs. Another study proposed the hSef binds to activated MEK, and inhibits the dissociation of MEK-ERK complexes thus inhibiting nuclear translocation of ERK and subsequent phosphorylation of nuclear ERK targets such as Elk-1.[61] This interaction of hSef with MEK occurred after MEK activation by either FGF or EGF and this complex was shown to be retained on the membrane or Golgi apparatus. Other studies showed that in fibroblasts, hSef-a over expression inhibited FGF-induced cell proliferation in an ERK independent manner, but also inhibition of Akt activation and increased p38 MAPK activation.[62] However, in epithelial cells forced expression of hSef-a inhibited FGF stimulated ERK activation, and hSef-a knockdown increased ERK activation and cell proliferation.[62]

Gene targeting studies in the mouse show that loss of Sef function does not result in an overt phenotype. Mice are born healthy and in normal Mendelian ratios. Early studies revealed that loss of Sef resulted in defects in the development of the auditory system.[63] Subsequent studies revealed that Sef gene targeted mice have increased cortical bone thickness and increased proliferation and osteoblast differentiation of bone marrow stromal cells.[64] Several craniofacial and dwarfing syndromes that are caused by point mutations in FGFR1, FGFR2, and FGFR3 demonstrate the importance of FGF signaling to skeletal development and homeostasis.[8,65] These mutations are autosomal dominant gain-of-function mutations. Therefore, it is tempting to speculate that Sef may have a role to modulate FGF signaling during skeletal growth or remodeling. Further studies with conditional gene targeted mice will provide new insight into the role of Sef in modulating FGF signaling during development and postnatal growth, homeostasis, and injury repair.

Sef has been reported to be dysregulated in various types of cancer, primarily of epithelial origin.[66] Sef is highly expressed normal human epithelial tissues such as thyroid gland, prostate and breast. Sef was low or undetectable in 73% of breast tumors, 69% of thyroid tumors and 52% of prostate tumors. The FGF signaling pathway has been shown to be an important contributor to prostate tumorigenesis.[67] Forced expression of Sef in prostate cancer cell lines inhibited their proliferation, migration and invasion and FGF induced ERK phosphorylation and Elk1 luciferase reporter activity.[67] These data strongly suggest that Sef functions as a tumor suppressor in tumors of epithelial origin. Further investigation will be required to determine why other feedback inhibitors of FGF signaling such as Sprys do not compensate for loss of Sef.

4. Regulation by Dual-specificity Phosphatases

The MAPKs are major downstream components of the FGF signaling pathway and include extracellular-regulated kinases (ERK), p38 MAP kinase (MAPK) and c-Jun NH_2-terminal kinase (JNK). The activity and cellular localization of these kinases must be tightly regulated to insure proper cellular responses including migration, proliferation, differentiation and survival. The duration, intensity and localization of FGF-mediated MAPK signaling is determined in part by the dual-specificity phosphatases (DUSP), also known as MAPK phosphatases (MKP).[68,69] The DUSPs dephosphorylate both threonine and tyrosine residues on the MAPK that they target as substrates, thus negatively regulating MAPKs and attenuating MAPK signaling.

The DUSPs have a common conserved structure with an N-terminal domain that determines substrate-binding specificity through its kinase interacting motif (KIM) and a C-terminal catalytic domain.[68] The DUSPs catalytic domain contains a highly conserved sequence DX_{26}(V/L)X(V/I)HCXAG(I/V)-SRSXT(I/V)XXAY(L/I)M where X is any amino acid. The DUSPs target different MAPKs although there is some overlap in their specificities. The DUSPs are divided into subclasses based on target specificity and cellular localization.[68] The DUSPs have been extensively reviewed elsewhere,[68,69] and this chapter will focus DUSP6 regulation of FGF signaling.

DUSP6 is highly specific for ERK1/2 and does not bind or dephosphorylate other MAPKs.[70] DUSP6 shuttles between the cytoplasm and the nucleus, and is retained primarily in the cytoplasm due to a leucine-rich nuclear export motif in the NH_2-terminal domain.[71] FGF induced ERK activation is a major activator of DUSP6 expression, constituting a feedback loop on FGF signaling. DUSP6 transcription is activated within 30 minutes of FGF stimulation and is mediated in part by the Ets2 transcription factor.[72] Other transcriptional regulators of DUSP6 include NFκB, PBX1, Tcf-3 and β-catenin. Like the ERK pathway itself, DUSP6 is tightly regulated. The major mode of DUSP6 regulation is phosphorylation and degradation of the protein.[73] DUSP6 is a substrate for ERK *in vitro*,[73] and was also reported to be phosphorylated by the mammalian target of rapamycin (mTOR) pathway,[74] as well as casein kinase 2.[75] These phosphorylation events facilitate the proteasomal degradation of DUSP6 thus limiting feedback inhibition of FGF-ERK signaling.

Gene targeting and overexpression studies in several model organisms (*Drosophila*, *Xenopus*, zebrafish, chick and mouse) indicate that DUSP6 plays a crucial role in development. Anti-sense strategies in zebrafish development revealed a role for DUSP6 in axial patterning,[76] and anti-sense studies in the chick limb showed that DUSP6 has a role in attenuating apoptosis.[77] DUSP6 is expressed during mouse development in areas of active FGF signaling making it a member of the FGF synexpression group.[11,76] Gene targeting of DUSP6 in the mouse does not produce an embryonic lethal phenotype and may thus be compensated for by other DUSPs such as DUSP7 and DUSP9, which are spatially and temporally expressed in a pattern similar to DUSP6 (limb buds, brachial arches, midbrain).[78] Postnatally, DUSP6-/- mice show several abnormalities including dwarfism and craniosynostosis, phenotypes resembling mice expressing constitutively active forms of FGFR1 and FGFR2.[8] Interestingly, DUSP6-/- mice show growth plate abnormalities of the long bones similar to mice expressing a gain-of-function mutation in FGFR3.[79] Recently, a study using a chemical inhibitor of DUSP6 in zebrafish showed that increased FGF signaling lead to an increase in the cardiac progenitor cell pool at the expense of endothelial progenitors.[79] Similar results were obtained in DUSP6-/- mice that showed the absence of DUSP6 increased proliferation of cardiomyocytes, although this phenotype may be dependent upon genetic background.[80] These phenotypes suggest a defect in feedback regulation of FGF signaling. FGFs induce the activation of several downstream signaling pathways and these phenotypes are likely the result of dysregulation of ERK signaling. This notion is supported by transgenic mice expressing activating mutations in MEK1 and MEK2, which are activators of ERK1/2, which show a skeletal defects similar to those observed in DUSP6$^{-/-}$ mice and mice expressing constitutively activated FGFRs.[81] Further evidence that loss of DUSP6 increases basal ERK activity was demonstrated in mouse embryo fibroblasts isolated from DUSP6$^{-/-}$ mice.[80] Other RTK ligands such as PDGF, VEGF, hepatocyte growth factor (HGF) and EGF also induce transcription of DUSP6 in an immediate-early manner, although studies on the role of DUSP6 in regulating their respective pathways *in vivo* are lacking.

Similar to other members of the FGF synexpression group, DUSP6 is dysregulated in several types of cancer.[68] In pancreatic cancer and several pancreatic cancer cell lines, the DUSP6 gene is often under expressed due to hypermethylation of its promoter.[82] Adenoviral expression of DUSP6 in these cell lines resulted in decreased ERK phosphorylation, reduced cell proliferation and increased apoptosis. Down regulation of DUSP6 expression has also been reported in lung cancer and reintroduction of DUSP6 in these cells resulted reduced cell growth.[83] In other types of cancer, DUSP6 is upregulated due to activating mutations in components of the Ras-ERK pathway.[84] In melanoma, where B-Raf is frequently mutated (V600E) MEK and ERK are hyperactivated.[85] Melanoma tumors and cell lines with this mutation showed an increase in expression of DUSP6 likely a result of increased ERK activity. Increased expression of DUSP6 has also been observed in breast cancers with oncogenic mutations in H-Ras. DUSP6 is also up regulated in gliomas with activating mutations in EGFR.[86] In each of these tumor types, overactivation of the MEK-ERK pathway results in increased transcriptional activation of the DUSP6 gene. Further study will be required to determine whether DUSP6 will be a suitable target for cancer therapeutics.

5. Regulation by FLRT Gene Family

The fibronectin-leucine-rich-transmembrane (FLRT) gene family was identified by screening a human skeletal muscle cDNA library with a degenerate oligonucleotide probe designed to the G2 domains of nidogen.[87] Three FLRT family members were identified, and shown to be type I transmembrane proteins with an extracellular domain consisting of ten leucine-rich repeats flanked by cysteine-rich domains, a fibronectin-like domain, a transmembrane domain and a short cytoplasmic domain. Comparison of mouse and human FLRT1, FLRT2, and FLRT3 sequences shows that there is a very high level of conservation of nucleotide and amino acid sequences. Mouse and human FLRT1 are 96% similar in amino acid sequence, FLRT2 97% similar, and FLRT3 96% similar.[87] During mouse development, FLRT1 is expressed primarily in the midbrain-hindbrain boundary, trigeminal and dorsal root ganglia, parts of the eye, and the foregut from E9-11.[88] FLRT2 is expressed in cephalic mesenchyme around E9.5-11, and a striped pattern along the trunk in a region of the sclerotome at E10.5, and at E11 in the brachial arches and limb buds.[88] FLRT3 was expressed in the developing brain including the midbrain and the forebrain-midbrain boundary, somites, and regions surrounding the heart from E9.5-10.5. At E10.5 expression was observed in the apical ectodermal ridge, mesoderm of the head and brachial arches.[88] Due to its expression pattern, FLRT3 is considered to be a member of the FGF synexpression group.[11]

Work by Bottcher and colleagues first demonstrated a role for FLRT3 in regulating FGF signaling.[89] High throughput *in situ* hybridization screening identified *Xenopus* FLRT3 (XFLRT3) as a gene with an expression pattern that was similar to FGF8. RT-PCR analysis showed that injection of mRNA encoding a dominant negative FGFR (XFD) into *Xenopus* embryos resulted in down regulation of XFLRT3 expression.[89] Conversely, injection of mRNA encoding either eFGF or FGF8 induced expression of XFLRT3 in *Xenopus* animal cap assays. Further analysis of XFLRT3 function revealed that injection of XFLRT3 mRNA into *Xenopus* embryos resulted in microcephaly and ectopic tails similar to the phenotype produced by injection of constitutively active FGFR1 thus mimicking activation of the FGF pathway.[89] This result was phenocopied by injecting mRNA encoding human FLRT2 and FLRT3. Structure-function analysis revealed that expression of the intracellular domain was sufficient to induce the phenotype. Co-immunoprecipitation experiments showed that XFLRT3 physically associated with FGFR1 and that this interaction required the fibronectin type III domain.[89] Subsequent studies show that mouse FLRT1, FLRT2, and FLRT3 all co-immunoprecipitated with FGFR1 when over expressed in 293T cells.[88] This association appeared to be constitutive as these complexes were formed without exogenous FGF-2 and the addition of FGF-2 did not increase the association between FLRT and FGFR1. Interestingly, these studies also showed that FLRTs interacted with an FGFR1 fusion protein consisting of a constitutively active kinase and transmembrane domain of FGFR1 and immunoglobulin Fc calling into question the requirement for the FNIII domain of FLRT in the interaction with the extracellular domain of FGFR1. In a subsequent study using co-immunoprecipitation, GST-fusion proteins and a yeast 2-hybrid approach it was show that both the intracellular and extracellular domains of FLRT2 were able to bind to FGFR2.[90] Furthermore, it was shown that the leucine-rich repeats were necessary and the FNIII domain was dispensable for the interaction of the extracellular domain with FGF-2. Although several studies support a physical interaction between FLRTs and FGFRs, and this interaction increases ERK signaling, a more systematic analysis of the interactions between all FLRTs and all FGFRs may alleviate the current discrepancies in the domains necessary for FLRT-FGFR interactions.

While ample evidence exist for an interaction between FGFRs and FLRTs and that this interaction may potentiate FGF-mediated ERK signaling, the mechanisms of this interaction remain poorly understood. A recent study showed that FLRT1 was tyrosine phosphorylated on its cytoplasmic domain in and FGFR1 and c-Src-dependent manner.[91] Evidence from this work suggests that FLRT1 is not a direct substrate of c-Src, but that c-Src potentiates the kinase activity of FGFR1 towards FLRT1. Overexpression of FLRT1 resulted in increased FGF ligand-dependent activation of ERK phosphorylation.

However, a mutant of FLRT1 (FLRT1-Y3F) where three tyrosines in the cytoplasmic domain had been mutated phenylalanine rendering it unable to be phosphorylated by FGFR1, resulted in chronic ERK activation in the absence of FGF.[91] Interestingly, this activity of FLRT1-Y3F required FGFR1 and c-Src, suggesting the mutant FLRT1 is a constitutive activator of FGFR1 and ERK activity. FLRTs have also been shown to function as regulators of homotypic cell adhesion. In other cases, FLRTs inhibit cell adhesion, and in both cases seem to be independent of FGFR signaling.[92] These discrepancies are likely to be cell-type specific and regulated by other proteins. The dual role of FLRTs as regulators of cell adhesion and FGF signaling is not unique. The homophilic cell adhesion molecules NCAM and N-cadherin also interact with and regulate FGFR signaling (see below). Further study will be required to determine whether the regulation of FGFR activity and cell adhesion by FLRTs are independent or coordinated events and the mechanisms by which this occurs.

6. Regulation by FGFRL1

Fibroblast growth factor receptor like 1 (FGFRL1) contains three extracellular immunoglobulin-like domains and a transmembrane domain with similarity to the FGFR family of receptor tyrosine kinases, but it lacks a tyrosine kinase domain and instead has a short cytoplasmic tail.[93] FGFRL1 was originally described by three groups and was also called FGFR5.[94–96] FGFRL1 is found in vertebrates and some non-chordates, however it is not found in *Drosophila* or *C. elegans.* During mouse development, FGFRL1 is highly expressed in cartilaginous structures such as trachea, ribs, nasal cartilage, vertebrae, and some muscles of the tongue and diaphragm.[97] FGFRL1 is also expressed at moderate levels in heart, lung, aorta, liver, kidney, and tendon, and much lower levels in other tissues.

FGFRL1 is a type I transmembrane protein with an extracellular domain that shares 39–42% sequence similarity to the extracellular domain of classical FGFRs.[98] The extracellular domain consists of a signal peptide, three Ig-like domains designated D1, D2 and D3 and a linker region called the “acidic box” found in conventional FGFRs located between Ig-loop D1 and D2.[98] The D2 domain of FGFRL1 contains a stretch of very basic residues analogous to the D2 domain of conventional FGFRs that has been shown to bind heparin. The most highly conserved domain between FGFRL1 and FGFRs is the D3 domain, which in FGFRs is part of the ligand-binding domain. Human FGFRL1 contains four N-linked glycosylation sites that are highly conserved among species. Three of these four sites are glycosylated as determined indirectly by mass spectrometry analysis. The intracellular domain of FGFRL1 is 100 amino acids in length and lacks the tyrosine kinase domain found on conventional FGFRs. The cytoplasmic domain is the least conserved among vertebrate FGFRL1 sequences. The exception is a stretch of 10 histidine residues alternating with other residues in the extreme C-terminus. This region has been shown to bind zinc and nickel ions.[99] The C-terminus also contains a tandem-based tyrosine motif that is similar to a tandem tyrosine motif that is a signal for trafficking of FGFRL1 from the plasma membrane to endosomes.[100] When this tyrosine motif was mutated, FGFRL1 trafficking was disrupted and its residence time in the membrane was increased. Shedding of the extracellular domain of FGFRL1 from the plasma membrane was first observed in cultured skeletal muscle myoblasts and myotubes.[101] The cleavage site for shedding of human FGFRL1 was mapped Ser369 and Ser370 and a secondary site at Pro361 and Gln362. The protease responsible of cleavage and shedding has not been identified.

Due to the sequence similarity between FGFRL1 and conventional FGFRs, and because FGFRL1 lacks a tyrosine kinase domain, it has been proposed that it is a regulator of FGF signaling functioning as either a dominant negative FGFR or as a decoy receptor.[93] As a decoy receptor, FGFRL1 could serve its function in either a membrane bound or soluble state. Overexpression of FGFRL1 is able to inhibit an FGF-responsive luciferase reporter in cultured cells.[100] Indeed, studies now show that FGFRL1 binds FGF2, FGF3, FGF4, FGF8, FGF10 and FGF22 as determined by dot blot assay, surface plasmon resonance, and cellular binding assays.[101] Other FGFs either did not bind FGFRL1 or interacted very weakly. FGFRL1 also interacts with heparin as determined by its high affinity interaction with immobilized heparin.[102] This interaction was stronger than the interaction of heparin with conventional FGFRs as determined by the concentration NaCl required for elution from immobilized heparin; 300–400 mM NaCl for conventional FGFRs and 600–700 mM NaCl for FGFRL1.

Whether FGFRL1 modulates FGFR signaling or has other functions independent of regulation of FGF signaling will require further study. FGFRL1 has pleotropic effects in cell culture. FGFRL1 has been shown to inhibit cell proliferation and promote differentiation. Over-expression of FGFRL1 in MG63 osteosarcoma cells and human embryonic 293 T cells inhibited cell proliferation and increased apoptosis.[98,103] The notion that FGFRL1 promotes differentiation was first proposed when it was observed that upon induction of skeletal muscle differentiation of C2C12 cells or primary myoblasts, FGFRL1 expression was strongly upregulated and thus correlated with the differentiation status of the cells.[103] FGFRL1 also has been shown to play a role in cell adhesion. The soluble, extracellular form of FGFRL1 when coated on plastic dishes promoted adhesion of various cell lines including 293T, CHO and murine 3T3 cells.[102] Interestingly, a soluble extracellular form of FGFR1 did not promote cell adhesion. It is possible that the adhesive properties of FGFRL1 are mediated by cell surface heparan sulfate proteoglycans. This notion is supported by evidence that soluble heparin blocked FGFRL1 mediated cell adhesion or by a mutation in the putative heparin binding domain of Ig-loop D2.[102] FGFRL1 may also play a role in cell-cell fusion as demonstrated by overexpression of FGFRL1 in CHO cells, which resulted in syncytia formation with multiple nuclei. Mutagenesis experiments showed that truncation of the cytoplasmic domain of FGFRL1 enhanced syncytia formation and that Ig-loop D3 and the transmembrane domain were necessary for cell-cell fusion.[103] Because FGFRL1 expression increases during myoblast differentiation and fusion into myotubes, it is possible that FGFRL1 promotes this process.

The *in vivo* function of FGFRL1 has been studied in several species including zebrafish, mouse and *Xenopus*. Injection of anti-sense morpholino oligonucleotides into zebrafish embryos cause defects in cartilage development and this is consistent with high-level expression of FGFRL1 in cartilage elements during mouse development.[104] In *Xenopus*, injection of mouse or human FGFRL1 mRNA in early stage blastomeres resulted in ventral and posterior truncations of *Xenopus* tadpoles, a phenotype very similar to that seen with overexpression of a dominant negative FGFR in *Xenopus* embryos.[101] Indeed, co-injection of FGFR1 mRNA with FGFRL1 mRNA rescued the phenotype induced by FGFRL1 alone. These experiments are strong evidence in support of a role for FGFRL1 in attenuating FGFR signaling. Injection of mRNA encoding the soluble form of FGFRL1 produced the same phenotype as full length FGFRL1 suggesting that it functions as a decoy receptor rather than as a dominant negative receptor.

Gene targeting studies in the mouse in which FGFRL1 expression was eliminated produced a neonatal lethal phenotype due to respiratory defects.[105] This defect was due to an underdeveloped diaphragm. Interestingly, other muscles appeared to be unaffected. FGFRL1 null mice also exhibited craniofacial abnormalities similar to mouse models of craniosynostosis where activating mutations in FGFRs were knocked in. Another profound phenotype seen in FGFRL1 gene targeted mice is agenesis of the metanephric kidney.[106] FGFRL1 is normally expressed in metanephric mesenchyme and renal vesicles. At E10.5 of mouse development, the ureteric bud invades the metanephric mesenchyme, undergoes branching and induces epithelial-mesenchymal transition (EMT) in the adjacent mesenchyme. In mutant FGFRL1 mice, the ureteric bud fails to branch correctly and the adjacent mesenchyme does not undergo EMT. Thus, FGFRL1 seems necessary for induction of metanephric mesenchyme. Interestingly, other organs that develop by branching morphogenesis such as the lung appear unaffected.

7. Regulation of FGF Signaling by Neural Cell Adhesion Molecules

Neural cell adhesion molecule (NCAM) is a member of a large family of cell adhesion molecules (CAMs) that contain both Ig-like domains and fibronectin type III repeats (FNIII) in their extracellular domains.[3,107] NCAM is comprised of five Ig-like domains and two FNIII repeats. NCAM is expressed as three isoforms that arise from alternative splicing of 27 exons. Two isoforms are transmembrane proteins, NCAM-A (180kDa) and NCAM-B (140 kDa) and vary only in the length of the cytoplasmic domain, and the third isoform, NCAM-C is a glycosylphosphatidylinositol-linked cell surface protein of 120kDa.[108] Another member of the CAM family, L1, has 6 Ig-like domains, and 5 to 6 FNIII repeats.[108]

NCAM mediates both homophilic and heterophilic cell-cell interactions in a calcium independent manner. The Ig-like domains mediate the homophilic binding (NCAM-NCAM), whereas the FNIII repeats mediate heterophilic interactions between NCAMs and other cell surface molecules, the most well characterized of which is FGFR.[108] In addition to mediating cell-cell interactions, CAMs function as cell signaling receptors. CAMs are abundantly expressed in the nervous system. NCAM plays a role in cell adhesion and neurite outgrowth, which requires both NCAM-NCAM interactions and NCAM-FGFR interactions, which initiate distinct intracellular signaling cascades.[108] Optical biosensor (Biacore) studies using recombinant proteins derived from various NCAM Ig and FNIII domains and the Ig domains of FGFR1 indicate direct interactions between FNIII domains 2 and 3 of NCAM and D2 and D3 of FGFR1.[109] Co-immunoprecipitation experiments suggest that D1, D2 and the linker region (acidic box) interact with NCAM.[110] The reasons for these differences are not known and likely are a reflection of the different experimental approaches used.

NCAM and L1 are capable of activating FGFR signaling without exogenous FGF ligands. Early studies showed that NCAM and L1 induced neurite outgrowth was inhibited in PC12 cells by a dominant negative FGFR suggesting that NCAM-FGFR interactions activate FGFR.[111] A synthetic peptide derived from the FG loop of the second FNIII domain of NCAM is sufficient to activate FGFR and induce tyrosine phosphorylation without exogenous FGF ligand in PC12 cells and primary neurons. In another study, it was shown that a FNIII derived peptide inhibited FGF-mediated FGFR activation, possibly through competition for the FGF binding site.[112] Yet another study using various isoforms of NCAM and FGFR1IIIc suggested that NCAM and FGF bind to separate sites on FGFR.[110] A peptide from the C-terminal region of first FNIII repeat called FGL and a peptide from the second FNIII repeat termed BCL bind to FGFR indicating that there are multiple FGFR binding sites on NCAM. In addition, FNIII repeats I-IV of L1 also activate FGFRs leading to receptor phosphorylation and neurite extension. Together, these investigations suggest a complex regulatory interaction between NCAM and FGFR.

While the interaction between NCAM and FGFR was originally studied in the context of neural cell adhesion and neurite extension, the association between FGFR and NCAM has been demonstrated in several types of non-neural cells including fibroblasts, epithelial and tumor cells. Studies using a soluble form of NCAM or the FGL peptide in HeLa cells, which are devoid of NCAM, showed that these two ligands induced tyrosine phosphorylation of FGFR1 but not FGFR2, FGFR3 or FGFR4 all of which are expressed in HeLa cells.[113] Thus, NCAM ligands preferentially activate FGFR1 in HeLa cells. These studies also revealed that FGF2 and NCAM ligands activate distinct FGFR signaling pathways. FGF2 and NCAM or FGL activated tyrosine phosphorylation of PLCg and FRS2 to the same extent, whereas NCAM and FGL activated c-Src and FGF2 did not.[113] Conversely, FGF2 induced tyrosine phosphorylation of Shc and NCAM and FGL did not. ERK was activated to a similar extent by FGF2, NCAM and FGL. ERK activation by NCAM and FGL stimulation was mediated by c-Src, whereas ERK activation by FGF2 was Ras-dependent. A chemical inhibitor of FGFRs, PD173074, inhibited signaling induced by either FGF-2 or NCAM ligands supporting the conclusion that NCAM and FGF-2 activate distinct FGFR1-dependent pathways. In addition, in time course studies FGF-2 stimulation induced only transient phosphorylation of FRS2, Akt, and c-Src, however Shc and ERK phosphorylation were sustained. In contrast, NCAM ligand stimulation induced sustained phosphorylation of FRS2, Akt and Src, with little phosphorylation of Shc and only transient activation of ERK. Immunofluorescent microscopy studies revealed that FGF-2 and

NCAM ligands induced differential trafficking of FGFR1. Stimulation with NCAM ligands results in recycling of FGFR1 to the cell surface, whereas, FGF-2 induced FGFR1 degradation via the lysosomal pathway. These studies also showed that NCAM ligands induced FGFR1 dependent cell migration but not proliferation, which correlated with FGFR1 recycling to the cell surface.[113] Conversely, FGF-2 induced FGFR1 dependent proliferation but not cell migration, which correlated with receptor degradation. Evidence suggests that differences in FGFR1 trafficking induced by NCAM or FGF2 lie in the inability of NCAM to induce recruitment of the E3 ubiquitin ligase c-Cbl to the FGFR1/FRS2 complex, which is required for FGFR1 degradation. In the absence recruitment of c-Cbl to FRS2, FGFR1 is recycled to the plasma membrane. In addition, evidence suggests that c-Src prevents the association of c-Cbl with FRS2, although the mechanisms by which this occurs requires further investigation. Indeed, preventing FGFR1 ubiquitination and degradation by expression of a dominant negative c-Cbl resulted in conferring migratory activity to FGF-2 due to increased FGFR1 stability via Rab11-dependent receptor recycling.[113] More recently studies have extended these observations by showing that activation of FRS2a by the FGL peptide was weaker than activation by FGF1[114] suggesting differential activation of signaling downstream of FGFR1 by NCAM and FGF.

8. Regulation of FGF Signaling by Cadherins

Cadherins are type I transmembrane proteins that mediate calcium-dependent homophilic binding in the formation of cell-cell adherens junctions.[115] The extracellular domain comprised of five calcium-binding cadherin repeats mediates the homophilic interactions. The short cytoplasmic domain with p120 catenin in the juxtamembrane region, and the C-terminal tail interacts with β- and γ-catenins. A link between cadherins and the cytoskeleton is achieved through the interaction of α-catenin with β-catenin and actin. Through these interactions, cadherins regulate cytoskeletal dynamics, β-catenin availability for Wnt signaling, activation of Rho GTPases, and regulate RTK signaling.[108,115] Cadherin family members are expressed in a cell type-specific manner, with some cell types expressing multiple cadherins. For example, endothelial cells express N-cadherin and VE-cadherin. E-cadherin is expressed predominantly in epithelial cells, whereas N-cadherin expression is more prevalent in mesenchymal cells. Indeed, cells undergoing epithelial to mesenchymal transition show a switch in expression from E-cadherin to N-cadherin.

N-cadherin is involved in regulating FGFR function and activity through a direct physical interaction. Binding of N-cadherin to FGFR1 occurs through acidic box region between Ig-loops D1 and D2 as determined in co-immunoprecipitation experiments with various deletion mutants of FGFR1.[110] Many biological functions mediated by N-cadherin are mediated through FGFR signaling. Chemical inhibitors of FGFR or expression of a dominant negative FGFR inhibits neurite outgrowth and guidance. Studies show that plating neural cells on recombinant N-cadherin leads to neurite extension without FGF-2, however inhibition of FGFR function abrogated neurite outgrowth indicating that process is FGFR dependent.[116] Other biological properties mediated by N-cadherin such as angiogenesis and tumor metastasis are also disrupted by FGFR inhibitors.[117] Studies suggest that in tumor cells, N-cadherin interaction with FGFR leads to prolonged membrane localization of FGFR resulting in sustained ERK activation and matrix metalloproteinase (MMP) production, thus facilitating invasion and metastasis.[118] Soluble N-cadherin stimulates endothelial ERK activation and cell migration in an FGFR dependent manner. Recent studies show that in endothelial cells, basal FGF-2 induced cell migration is promoted by N-cadherin and inhibited by VE-cadherin.[119] The inhibitory effect of VE-cadherin on endothelial cell migration was in part due to VE-cadherin association with FGFR together with density-enhanced phosphatase (Dep-1), which dephosphorylates FGFR and inhibits receptor signaling. Thus, cadherins can either be positive or negative regulators of FGF signaling depending upon the cellular context and the specific cadherins that are expressed.

9. Regulation of FGF Signaling by Integrins

Integrins link cells to the extracellular matrix (ECM) and are a means by which cells translate extracellular cues into intracellular signaling pathway integration. Integrins are transmembrane proteins that are comprised of one *alpha* and one *beta* sub-unit and form a non-covalent dimeric complex. There are eighteen *alpha* subunits and eight *beta* subunits that combinatorially form up to 24 unique integrin receptors with unique specificities for various ECM components.[3] Integrins have been shown to regulate FGF signaling at several levels. First, integrin $\alpha_5\beta_1$ which binds fibronectin, upregulated the expression of FGF-2.[120] Second, FGF-2 binds directly to integrin $\alpha_v\beta_3$ to stimulate cell adhesion, spreading and migration.[121] Furthermore, FGFR1 has been shown to physically associate with integrin $\alpha_v\beta_3$ in a complex that potentiates FGFR1 mediated ERK activation.[122] Recently, studies showed that fibronectin (FN) could activate FGFR1 in the absence of FGF ligand in liver endothelial cells.[123] Activation of FGFR1 by FN required integrin β_1 as determined by studies using siRNA and neutralizing antibodies. Further studies using chemical inhibitors and genetic approaches revealed that c-Src is required for FN mediated transactivation of FGFR1.[123] These studies also showed that FGF-induced phosphorylation of FGFR1 preferentially activate ERK signaling, whereas FN-mediated activation of FGFR1 preferentially activates Akt-Rac1 signaling and cell migration. Thus, FN can transactivate FGFR1 via integrin β_1-Src signaling in an FGF-ligand independent manner.

10. Summary and Perspective

This chapter has highlighted some of the many mechanisms by which FGFR signaling is regulated (Fig. 1). Cell adhesion molecules can positively or negatively regulate FGFR signaling in a context and cell type specific manner, and can be either FGF-ligand dependent or independent. Other transmembrane proteins such as FLRT3 positively influence FGFR signaling through interaction with its extracellular FNIII domain. Sef, and inhibitor of FGFR signaling also has a FNIII

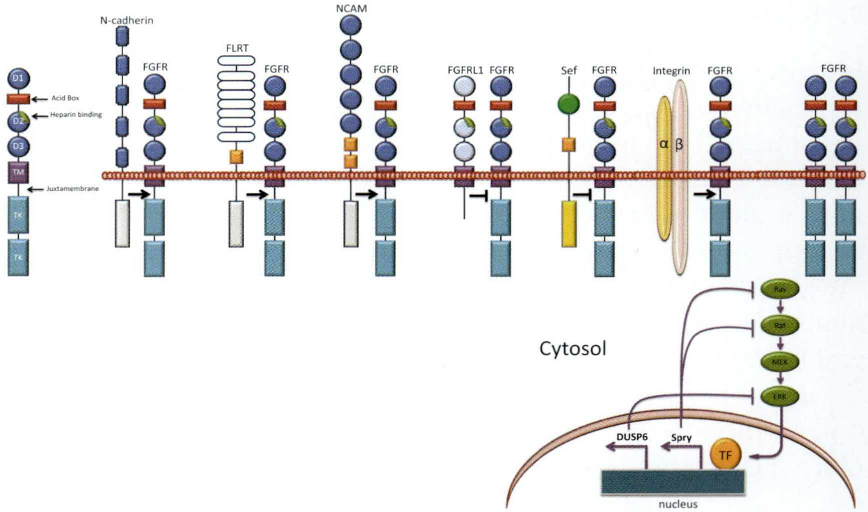

Fig. 1. Regulation of FGFR signaling pathways. A. The prototype FGF receptor is comprised of an extracellular domain, a transmembrane domain (TM) and a cytoplasmic domain consisting of a juxtamembrane region, and a split tyrosine kinase (TK) domain. The extracellular domain is made up of three immunoglobulin-like domains designated D1, D2 and D3. There is a highly conserved stretch of acidic amino acids called the acidic box located between D1 and D2. There is also a heparin-binding domain located in N-terminal half of D2. B. Canonical FGFR signaling is initiated by FGF ligand binding to its receptor resulting in dimerization and activation of the tyrosine kinase. FGFRs can also be activated (→) by several transmembrane proteins through protein-protein interactions in the absence of FGF ligand. Among these transmembrane activators of FGFR signaling are N-cadherin, members of the FLRT family, NCAM, and integrins. FGFR signaling can also be inhibited (—|) by interaction with specific transmembrane proteins including FGFRL1 and Sef. Cytoplasmic regulators also inhibit FGFR signaling. FGF signaling induces the expression of DUSP6, a dual specificity phosphatase, which acts as a feed back inhibitor by dephosphorylating and inactivating ERK. Another family of feed back inhibitors called Sprouty (Spry) are induced by FGF signaling and inhibit ERK signaling at or above the level of Ras and at the level of Raf. Thus there are multiple modes of positive and negative regulation of FGFR signaling.

repeat, and it is tempting to speculate that the inhibitory function of Sef may also be mediated in part through its FNIII domain. Prior studies show that the transmembrane domain of Sef is necessary for its inhibitory function on FGF signaling, but do not exclude a role for the FNIII domain in the inhibitory function of Sef. Due to the biological output of FGFR signaling is dependent on the intensity and duration of the FGF signal, multiple levels of regulation have apparently evolved to provide context specific fine-tuning of FGF signaling in a spatial and temporal dependent manner. Further study will be required to determine the hierarchical regulation of positive and negative inputs of FGFR signaling.

Acknowledgements

I apologize to my many colleagues whose work could not be cited due to space limitations. I thank the members of my laboratory, past and present, which have contributed to our knowledge of FGF signaling. This work was supported in part by NIH grant 5P30GM103392 to R. Friesel.

References

1. Goetz, R. and Mohammadi, M. Exploring mechanisms of FGF signalling through the lens of structural biology. *Nat Rev Mol Cell Biol* **14**, 166–180 (2013).
2. Korc, M. and Friesel, R.E. The role of fibroblast growth factors in tumor growth. *Curr Cancer Drug Targets* **9**, 639–651 (2009).
3. Murakami, M., Elfenbein, A. and Simons, M. Non-canonical fibroblast growth factor signalling in angiogenesis. *Cardiovasc Res* **78**, 223–231 (2008).
4. Beenken, A. and Mohammadi, M. The FGF family: Biology, pathophysiology and therapy. *Nat Rev Drug Discov* **8**, 235–253 (2009).
5. Hatch, N.E. FGF signaling in craniofacial biological control and pathological craniofacial development. *Crit Rev Eukaryot Gene Expr* **20**, 295–311 (2010).
6. Melville, H., Wang, Y., Taub, P.J. and Jabs, E.W. Genetic basis of potential therapeutic strategies for craniosynostosis. *Am J Med Genet A* **152A**, 3007–3015 (2010).
7. Scialla, J.J. and Wolf, M. Roles of phosphate and fibroblast growth factor 23 in cardiovascular disease. *Nat Rev Nephrol* **10**, 268–278 (2014).
8. Su, N., Du, X. and Chen, L. FGF signaling: its role in bone development and human skeleton diseases. *Front Biosci* **13**, 2842–2865 (2008).

9. Eswarakumar, V.P., Lax, I. and Schlessinger, J. Cellular signaling by fibroblast growth factor receptors. *Cytokine Growth Factor Rev* **16**, 139–149 (2005).
10. Thisse, B. and Thisse, C. Functions and regulations of fibroblast growth factor signaling during embryonic development. *Dev Biol* **287**, 390–402 (2005).
11. Tsang, M. and Dawid, I.B. Promotion and attenuation of FGF signaling through the Ras-MAPK pathway. *Sci STKE* **2004**, pe17 (2004).
12. Hacohen, N., Kramer, S., Sutherland, D., Hiromi, Y., *et al.* sprouty encodes a novel antagonist of FGF signaling that patterns apical branching of the Drosophila airways. *Cell* **92**, 253–263 (1998).
13. Minowada, G., *et al.* Vertebrate Sprouty genes are induced by FGF signaling and can cause chondrodysplasia when overexpressed. *Development* **126**, 4465–4475 (1999).
14. Edwin, F., Anderson, K., Ying, C. and Patel, T.B. Intermolecular interactions of Sprouty proteins and their implications in development and disease. *Mol Pharmacol* **76**, 679–691 (2009).
15. Mason, J.M., Morrison, D.J., Basson, M.A. and Licht, J.D. Sprouty proteins: Multifaceted negative-feedback regulators of receptor tyrosine kinase signaling. *Trends Cell Biol* **16**, 45–54 (2006).
16. Gross, I., Bassit, B., Benezra, M. and Licht, J.D. Mammalian sprouty proteins inhibit cell growth and differentiation by preventing ras activation. *J Biol Chem* **276**, 46460–46468 (2001).
17. Sasaki, A., *et al.* Mammalian Sprouty4 suppresses Ras-independent ERK activation by binding to Raf1. *Nat Cell Biol* **5**, 427–432 (2003).
18. Egan, J.E., Hall, A.B., Yatsula, B.A. and Bar-Sagi, D. The bimodal regulation of epidermal growth factor signaling by human Sprouty proteins. *Proc Natl Acad Sci USA* **99**, 6041–6046 (2002).
19. Lim, J., *et al.* Sprouty proteins are targeted to membrane ruffles upon growth factor receptor tyrosine kinase activation. Identification of a novel translocation domain. *J Biol Chem* **275**, 32837–32845 (2000).
20. Basson, M.A., *et al.* Sprouty1 Is a Critical Regulator of GDNF/RET-Mediated Kidney Induction. *Dev Cell* **8**, 229–239 (2005).
21. Chi, L., *et al.* Sprouty proteins regulate ureteric branching by coordinating reciprocal epithelial Wnt11, mesenchymal Gdnf and stromal Fgf7 signalling during kidney development. *Development* **131**, 3345–3356 (2004).
22. Ding, W., Bellusci, S., Shi, W. and Warburton, D. Functional analysis of the human Sprouty2 gene promoter. *Gene* **322**, 175–185 (2003).

23. Ding, W., Bellusci, S., Shi, W. and Warburton, D. Genomic structure and promoter characterization of the human Sprouty4 gene, a novel regulator of lung morphogenesis. *Am J Physiol Lung Cell Mol Physiol* **287**, L52–L59 (2004).
24. Lee, S.H., Schloss, D.J., Jarvis, L., Krasnow, M.A., *et al.* Inhibition of angiogenesis by a mouse sprouty protein. *J Biol Chem* **276**, 4128–4133 (2001).
25. Akbulut, S., *et al.* Sprouty proteins inhibit receptor-mediated activation of phosphatidylinositol-specific phospholipase C. *Mol Biol Cell* **21**, 3487–3496 (2010).
26. Chow, S.Y., Yu, C.Y. and Guy, G.R. Sprouty2 interacts with protein kinase C delta and disrupts phosphorylation of protein kinase D1. *J Biol Chem* **284**, 19623–19636 (2009).
27. Hanafusa, H., Torii, S., Yasunaga, T. and Nishida, E. Sprouty1 and Sprouty2 provide a control mechanism for the Ras/MAPK signalling pathway. *Nat Cell Biol* **4**, 850–858 (2002).
28. Lao, D.H., *et al.* A Src Homology 3-binding Sequence on the C Terminus of Sprouty2 Is Necessary for Inhibition of the Ras/ERK Pathway Downstream of Fibroblast Growth Factor Receptor Stimulation. *J Biol Chem* **281**, 29993–30000 (2006).
29. Fong, C.W., *et al.* Tyrosine phosphorylation of Sprouty2 enhances its interaction with c-Cbl and is crucial for its function. *J Biol Chem* **278**, 33456–33464 (2003).
30. Hall, A.B., *et al.* hSpry2 is targeted to the ubiquitin-dependent proteasome pathway by c-Cbl. *Curr Biol* **13**, 308–314 (2003).
31. Chandramouli, S., *et al.* Tesk1 interacts with Spry2 to abrogate its inhibition of ERK phosphorylation downstream of receptor tyrosine kinase signaling. *J Biol Chem* **283**, 1679–1691 (2008).
32. Leeksma, O.C., *et al.* Human sprouty 4, a new ras antagonist on 5q31, interacts with the dual specificity kinase TESK1. *Eur J Biochem* **269**, 2546–2556 (2002).
33. Tsumura, Y., Toshima, J., Leeksma, O.C., Ohashi, K., *et al.* Sprouty-4 negatively regulates cell spreading by inhibiting the kinase activity of testicular protein kinase. *Biochem J* **387**, 627–637 (2005).
34. Lao, D.H., *et al.* Direct binding of PP2A to Sprouty2 and phosphorylation changes are a prerequisite for ERK inhibition downstream of fibroblast growth factor receptor stimulation. *J Biol Chem* **282**, 9117–9126 (2007).

35. Haglund, K., Schmidt, M.H., Wong, E.S., Guy, G.R., *et al.* Sprouty2 acts at the Cbl/CIN85 interface to inhibit epidermal growth factor receptor downregulation. *EMBO Rep* **6**, 635–641 (2005).
36. Impagnatiello, M.A., *et al.* Mammalian sprouty-1 and -2 are membrane-anchored phosphoprotein inhibitors of growth factor signaling in endothelial cells. *J Cell Biol* **152**, 1087–1098 (2001).
37. Aranda, S., Alvarez, M., Turro, S., Laguna, A., *et al.* Sprouty2-mediated inhibition of fibroblast growth factor signaling is modulated by the protein kinase DYRK1A. *Mol Cell Biol* **28**, 5899–5911 (2008).
38. Yang, X., *et al.* Sprouty genes are expressed in osteoblasts and inhibit fibroblast growth factor-mediated osteoblast responses. *Calcif Tissue Int* **78**, 233–240 (2006).
39. Nadeau, R.J., Toher, J.L., Yang, X., Kovalenko, D., *et al.* Regulation of Sprouty2 stability by mammalian Seven-in-Absentia homolog 2. *J Cell Biochem* **100**, 151–160 (2007).
40. Qi, J., *et al.* The ubiquitin ligase Siah2 regulates tumorigenesis and metastasis by HIF-dependent and -independent pathways. *Proc Natl Acad Sci USA* **105**, 16713–16718 (2008).
41. Mason, J.M., *et al.* Tyrosine phosphorylation of sprouty proteins regulates their ability to inhibit growth factor signaling: a dual feedback loop. *Mol Biol Cell* **15**, 2176–2188 (2004).
42. Hanafusa, H., Torii, S., Yasunaga, T., Matsumoto, K., *et al.* Shp2, an SH2-containing protein tyrosine phosphatase, positively regulates receptor tyrosine kinase signaling by dephosphorylating and inactivating the inhibitor sprouty. *J Biol Chem* **279**, 22992–22995 (2004).
43. Rubin, C., Zwang, Y., Vaisman, N., Ron, D., *et al.* Phosphorylation of carboxyl-terminal tyrosines modulates the specificity of Sprouty-2 inhibition of different signaling pathways. *J Biol Chem* **280**, 9735–9744 (2005).
44. DaSilva, J., Xu, L., Kim, H.J., Miller, W.T., *et al.* Regulation of sprouty stability by Mnk1-dependent phosphorylation. *Mol Cell Biol* **26**, 1898–1907 (2006).
45. Lim, J., *et al.* The cysteine-rich sprouty translocation domain targets mitogen-activated protein kinase inhibitory proteins to phosphatidylinositol 4,5-bisphosphate in plasma membranes. *Mol Cell Biol* **22**, 7953–7966 (2002).
46. Cabrita, M.A., Jaggi, F., Widjaja, S.P. and Christofori, G. A Functional Interaction between Sprouty Proteins and Caveolin-1. *J Biol Chem* **281**, 29201–2912 (2006).

47. Ozaki, K., Miyazaki, S., Tanimura, S. and Kohno, M. Efficient suppression of FGF-2-induced ERK activation by the cooperative interaction among mammalian Sprouty isoforms. *J Cell Sci* **118**, 5861–5871 (2005).
48. Taketomi, T., *et al.* Loss of mammalian Sprouty2 leads to enteric neuronal hyperplasia and esophageal achalasia. *Nat Neurosci* **8**, 855–857 (2005).
49. Taniguchi, K., *et al.* Sprouty2 and Sprouty4 are essential for embryonic morphogenesis and regulation of FGF signaling. *Biochem Biophys Res Commun* **352**, 896–902 (2007).
50. Taniguchi, K., *et al.* Sprouty4 deficiency potentiates Ras-independent angiogenic signals and tumor growth. *Cancer Sci* **100**, 1648–1654 (2009).
51. Klein, O.D., *et al.* Sprouty genes control diastema tooth development via bidirectional antagonism of epithelial-mesenchymal FGF signaling. *Dev Cell* **11**, 181–190 (2006).
52. Shim, K., Minowada, G., Coling, D.E. and Martin, G.R. Sprouty2, a mouse deafness gene, regulates cell fate decisions in the auditory sensory epithelium by antagonizing FGF signaling. *Dev Cell* **8**, 553–564 (2005).
53. Yang, X., *et al.* Spry1 and Spry4 differentially regulate human aortic smooth muscle cell phenotype via Akt/FoxO/myocardin signaling. *PLoS One* **8**, e58746 (2013).
54. Tsang, M., Friesel, R., Kudoh, T. and Dawid, I.B. Identification of Sef, a novel modulator of FGF signalling. *Nat Cell Biol* **4**, 165–169 (2002).
55. Furthauer, M., Lin, W., Ang, S.L., Thisse, B., *et al.* Sef is a feedback-induced antagonist of Ras/MAPK-mediated FGF signalling. *Nat Cell Biol* **4**, 170–174 (2002).
56. Yang, R.B., *et al.* A Novel Interleukin-17 Receptor-like Protein Identified in Human Umbilical Vein Endothelial Cells Antagonizes Basic Fibroblast Growth Factor-induced Signaling. *J Biol Chem* **278**, 33232–33238 (2003).
57. Kovalenko, D., *et al.* A role for extracellular and transmembrane domains of Sef in Sef-mediated inhibition of FGF signaling. *Cell Signal* **18**, 1958–1966 (2006).
58. Xiong, S., *et al.* hSef inhibits PC-12 cell differentiation by interfering with Ras-mitogen-activated protein kinase MAPK signaling. *J Biol Chem* **278**, 50273–50282 (2003).
59. Kovalenko, D., Yang, X., Nadeau, R.J., Harkins, L.K., *et al.* Sef inhibits fibroblast growth factor signaling by inhibiting FGFR1 tyrosine phosphorylation and subsequent ERK activation. *J Biol Chem* **278**, 14087–14091 (2003).

60. Preger, E,. *et al.* Alternative splicing generates an isoform of the human Sef gene with altered subcellular localization and specificity. *Proc Natl Acad Sci USA* **101**, 1229–1234 (2004).
61. Torii, S., Kusakabe, M., Yamamoto, T., Maekawa, M., *et al.* Sef is a spatial regulator for Ras/MAP kinase signaling. *Dev Cell* **7**, 33–44 (2004).
62. Ziv, I., *et al.* The human Sef-a isoform utilizes different mechanisms to regulate FGFR signaling pathways and subsequent cell fate. *J Biol Chem* **281**, 39225–39235 (2006).
63. Abraira, V.E., *et al.* Changes in Sef levels influence auditory brainstem development and function. *J Neurosci* **27**, 4273–4282 (2007).
64. He, Q., *et al.* Deficiency of Sef is associated with increased postnatal cortical bone mass by regulating Runx2 activity. *Journal of bone and mineral research: the official journal of the American Society for Bone and Mineral Research* **29**, 1217–1231 (2014).
65. Katoh, M. FGFR2 abnormalities underlie a spectrum of bone, skin, and cancer pathologies. *J Invest Dermatol* **129**, 1861–1867 (2009).
66. Darby, S., *et al.* Loss of Sef (similar expression to FGF) expression is associated with high grade and metastatic prostate cancer. *Oncogene* **25**, 4122–4127 (2006).
67. Darby, S., *et al.* Similar expression to FGF (Sef) inhibits fibroblast growth factor-induced tumourigenic behaviour in prostate cancer cells and is downregulated in aggressive clinical disease. *British journal of cancer* **101**, 1891–1899 (2009).
68. Bermudez, O., Pages, G. and Gimond, C. The dual-specificity MAP kinase phosphatases: critical roles in development and cancer. *Am J Physiol Cell Physiol* **299**, C189–C202 (2010).
69. Caunt, C.J. and Keyse, S.M. Dual-specificity MAP kinase phosphatases (MKPs): shaping the outcome of MAP kinase signalling. *FEBS J* **280**, 489–504 (2013).
70. Arkell, R.S., *et al.* DUSP6/MKP-3 inactivates ERK1/2 but fails to bind and inactivate ERK5. *Cell Signal* **20**, 836–843 (2008).
71. Karlsson, M., Mathers, J., Dickinson, R.J., Mandl, M., *et al.* Both nuclear-cytoplasmic shuttling of the dual specificity phosphatase MKP-3 and its ability to anchor MAP kinase in the cytoplasm are mediated by a conserved nuclear export signal. *J Biol Chem* **279**, 41882–41891 (2004).
72. Ekerot, M., *et al.* Negative-feedback regulation of FGF signalling by DUSP6/MKP-3 is driven by ERK1/2 and mediated by Ets factor binding to a conserved site within the DUSP6/MKP-3 gene promoter. *Biochem J* **412**, 287–298 (2008).

73. Marchetti, S., *et al.* Extracellular signal-regulated kinases phosphorylate mitogen-activated protein kinase phosphatase 3/DUSP6 at serines 159 and 197, two sites critical for its proteasomal degradation. *Mol Cell Biol* **25**, 854–864 (2005).
74. Bermudez, O., Marchetti, S., Pages, G. and Gimond, C. Post-translational regulation of the ERK phosphatase DUSP6/MKP3 by the mTOR pathway. *Oncogene* **27**, 3685–3691 (2008).
75. Castelli, M., *et al.* MAP kinase phosphatase 3 (MKP3) interacts with and is phosphorylated by protein kinase CK2alpha. *J Biol Chem* **279**, 44731–4449 (2004).
76. Tsang, M., *et al.* A role for MKP3 in axial patterning of the zebrafish embryo. *Development* **131**, 2769–2779 (2004).
77. Kawakami, Y., *et al.* MKP3 mediates the cellular response to FGF8 signalling in the vertebrate limb. *Nat Cell Biol* **5**, 513–519 (2003).
78. Li, C., Scott, D.A., Hatch, E., Tian, X., *et al.* Dusp6 (Mkp3) is a negative feedback regulator of FGF-stimulated ERK signaling during mouse development. *Development* **134**, 167–176 (2007).
79. Brodie, S.G. and Deng, C.X. Mouse models orthologous to FGFR3-related skeletal dysplasias. *Pediatr Pathol Mol Med* **22**, 87–103 (2003).
80. Maillet, M., *et al.* DUSP6 (MKP3) null mice show enhanced ERK1/2 phosphorylation at baseline and increased myocyte proliferation in the heart affecting disease susceptibility. *J Biol Chem* **283**, 31246–31255 (2008).
81. Murakami, S., *et al.* Constitutive activation of MEK1 in chondrocytes causes Stat1-independent achondroplasia-like dwarfism and rescues the Fgfr3-deficient mouse phenotype. *Genes Dev* **18**, 290–305 (2004).
82. Xu, S., Furukawa, T., Kanai, N., Sunamura, M., *et al.* Abrogation of DUSP6 by hypermethylation in human pancreatic cancer. *J Hum Genet* **50**, 159–167 (2005).
83. Okudela, K., *et al.* Down-regulation of DUSP6 expression in lung cancer: its mechanism and potential role in carcinogenesis. *Am J Pathol* **175**, 867–881 (2009).
84. Warmka, J.K., Mauro, L.J. and Wattenberg, E.V. Mitogen-activated protein kinase phosphatase-3 is a tumor promoter target in initiated cells that express oncogensic Ras. *J Biol Chem* **279**, 33085–33092 (2004).
85. Bloethner, S., *et al.* Effect of common B-RAF and N-RAS mutations on global gene expression in melanoma cell lines. *Carcinogenesis* **26**, 1224–1232 (2005).

86. Ramnarain, D.B., *et al.* Differential gene expression analysis reveals generation of an autocrine loop by a mutant epidermal growth factor receptor in glioma cells. *Cancer Res* **66**, 867–874 (2006).
87. Lacy, S.E., Bonnemann, C.G., Buzney, E.A. and Kunkel, L.M. Identification of FLRT1, FLRT2, and FLRT3: a novel family of transmembrane leucine-rich repeat proteins. *Genomics* **62**, 417–426 (1999).
88. Haines, B.P., Wheldon, L.M., Summerbell, D., Heath, J.K., *et al.* Regulated expression of FLRT genes implies a functional role in the regulation of FGF signalling during mouse development. *Dev Biol* **297**, 14–25 (2006).
89. Bottcher, R.T., Pollet, N., Delius, H. and Niehrs, C. The transmembrane protein XFLRT3 forms a complex with FGF receptors and promotes FGF signalling. *Nat Cell Biol* **6**, 38–44 (2004).
90. Wei, K., Xu, Y., Tse, H., Manolson, M.F., *et al.* Mouse FLRT2 interacts with the extracellular and intracellular regions of FGFR2. *J Dent Res* **90**, 1234–1239 (2011).
91. Wheldon, L.M., *et al.* Critical role of FLRT1 phosphorylation in the interdependent regulation of FLRT1 function and FGF receptor signalling. *PLoS One* **5**, e10264 (2010).
92. Chen, X., Koh, E., Yoder, M. and Gumbiner, B.M. A protocadherin-cadherin-FLRT3 complex controls cell adhesion and morphogenesis. *PLoS One* **4**, e8411 (2009).
93. Trueb, B. Biology of FGFRL1, the fifth fibroblast growth factor receptor. *Cell Mol Life Sci* **68**, 951–964 (2011).
94. Wiedemann, M. and Trueb, B. Characterization of a novel protein (FGFRL1) from human cartilage related to FGF receptors. *Genomics* **69**, 275–279 (2000).
95. Kim, I., Moon, S., Yu, K., Kim, U., *et al.* A novel fibroblast growth factor receptor-5 preferentially expressed in the pancreas(1). *Biochim Biophys Acta* **1518**, 152–156 (2001).
96. Sleeman, M., *et al.* Identification of a new fibroblast growth factor receptor, FGFR5. *Gene* **271**, 171–182 (2001).
97. Trueb, B. and Taeschler, S. Expression of FGFRL1, a novel fibroblast growth factor receptor, during embryonic development. *Int J Mol Med* **17**, 617–620 (2006).
98. Trueb, B., Zhuang, L., Taeschler, S. and Wiedemann, M. Characterization of FGFRL1, a novel fibroblast growth factor (FGF) receptor preferentially expressed in skeletal tissues. *J Biol Chem* **278**, 33857–33865 (2003).

99. Zhuang, L., Karotki, A.V., Bruecker, P. and Trueb, B. Comparison of the receptor FGFRL1 from sea urchins and humans illustrates evolution of a zinc binding motif in the intracellular domain. *BMC Biochem* **10**, 33 (2009).
100. Rieckmann, T., Zhuang, L., Fluck, C.E. and Trueb, B. Characterization of the first FGFRL1 mutation identified in a craniosynostosis patient. *Biochim Biophys Acta* **1792**, 112–121 (2009).
101. Steinberg, F., *et al.* The FGFRL1 receptor is shed from cell membranes, binds fibroblast growth factors (FGFs), and antagonizes FGF signaling in Xenopus embryos. *J Biol Chem* **285**, 2193–2202 (2010).
102. Rieckmann, T., Kotevic, I. and Trueb, B. The cell surface receptor FGFRL1 forms constitutive dimers that promote cell adhesion. *Exp Cell Res* **314**, 1071–1081 (2008).
103. Steinberg, F., Gerber, S.D., Rieckmann, T. and Trueb, B. Rapid fusion and syncytium formation of heterologous cells upon expression of the FGFRL1 receptor. *J Biol Chem* **285**, 37704–37715 (2010).
104. Hall, C., Flores, M.V., Murison, G., Crosier, K., *et al.* An essential role for zebrafish Fgfrl1 during gill cartilage development. *Mech Dev* **123**, 925–940 (2006).
105. Baertschi, S., Zhuang, L. and Trueb, B. Mice with a targeted disruption of the Fgfrl1 gene die at birth due to alterations in the diaphragm. *FEBS J* **274**, 6241–6253 (2007).
106. Gerber, S.D., Steinberg, F., Beyeler, M., Villiger, P.M., *et al.* The murine Fgfrl1 receptor is essential for the development of the metanephric kidney. *Dev Biol* **335**, 106–119 (2009).
107. Kiselyov, V.V. NCAM and the FGF-receptor. *Adv Exp Med Biol* **663**, 67–79 (2010).
108. Polanska, U.M., Fernig, D.G. and Kinnunen, T. Extracellular interactome of the FGF receptor-ligand system: Complexities and the relative simplicity of the worm. *Dev Dyn* **238**, 277–293 (2009).
109. Kiselyov, V.V., *et al.* Structural basis for a direct interaction between FGFR1 and NCAM and evidence for a regulatory role of ATP. *Structure* **11**, 691–701 (2003).
110. Sanchez-Heras, E., Howell, F.V., Williams, G. and Doherty, P. The fibroblast growth factor receptor acid box is essential for interactions with N-cadherin and all of the major isoforms of neural cell adhesion molecule. *J Biol Chem* **281**, 35208–35216 (2006).
111. Williams, E.J., Furness, J., Walsh, F.S. and Doherty, P. Activation of the FGF receptor underlies neurite outgrowth stimulated by L1, N-CAM, and N-cadherin. *Neuron* **13**, 583–594 (1994).

112. Francavilla, C., *et al.* Neural cell adhesion molecule regulates the cellular response to fibroblast growth factor. *J Cell Sci* **120**, 4388–4394 (2007).
113. Francavilla, C., *et al.* The binding of NCAM to FGFR1 induces a specific cellular response mediated by receptor trafficking. *J Cell Biol* **187**, 1101–1116 (2009).
114. Chen, Y., Li, S., Berezin, V. and Bock, E. The fibroblast growth factor receptor (FGFR) agonist FGF1 and the neural cell adhesion molecule-derived peptide FGL activate FGFR substrate 2alpha differently. *J Neurosci Res* **88**, 1882–1889 (2010).
115. Gumbiner, B.M. Regulation of cadherin-mediated adhesion in morphogenesis. *Nat Rev Mol Cell Biol* **6**, 622–634 (2005).
116. Utton, M.A., Eickholt, B., Howell, F.V., Wallis, J., *et al.* Soluble N-cadherin stimulates fibroblast growth factor receptor dependent neurite outgrowth and N-cadherin and the fibroblast growth factor receptor co-cluster in cells. *J Neurochem* **76**, 1421–1430 (2001).
117. Hulit, J., *et al.* N-cadherin signaling potentiates mammary tumor metastasis via enhanced extracellular signal-regulated kinase activation. *Cancer Res* **67**, 3106–3116 (2007).
118. Hazan, R.B., Phillips, G.R., Qiao, R.F., Norton, L., *et al.* Exogenous expression of N-cadherin in breast cancer cells induces cell migration, invasion, and metastasis. *J Cell Biol* **148**, 779–790 (2000).
119. Giampietro, C., *et al.* Overlapping and divergent signaling pathways of N-cadherin and VE-cadherin in endothelial cells. *Blood* **119**, 2159–2170 (2012).
120. Klein, S., Bikfalvi, A., Birkenmeier, T.M., Giancotti, F.G., *et al.* Integrin regulation by endogenous expression of 18-kDa fibroblast growth factor-2. *J Biol Chem* **271**, 22583–22590 (1996).
121. Rusnati, M., Tanghetti, E., Dell'Era, P., Gualandris, A., *et al.* alphavbeta3 integrin mediates the cell-adhesive capacity and biological activity of basic fibroblast growth factor (FGF-2) in cultured endothelial cells. *Mol Biol Cell* **8**, 2449–2461 (1997).
122. Tanghetti, E., *et al.* Biological activity of substrate-bound basic fibroblast growth factor (FGF2): recruitment of FGF receptor-1 in endothelial cell adhesion contacts. *Oncogene* **21**, 3889–3897 (2002).
123. Zou, L., Cao, S., Kang, N., Huebert, R.C., *et al.* Fibronectin induces endothelial cell migration through beta1 integrin and Src-dependent phosphorylation of fibroblast growth factor receptor-1 at tyrosines 653/654 and 766. *J Biol Chem* **287**, 7190–7202 (2012).

Chapter 3
FRS2α: At the Center of FGF Signaling

Cong Wang*,‡, Wallace L. McKeehan† and Fen Wang†,‡

Abstract

Since its identification in 1990s, FRS2 (fibroblast growth factor receptor substrate 2) has been recognized as an essential adaptor protein for most signaling activities of the fibroblast growth factors (FGF). The FRS2 family has two members, FRS2α and FRS2β, which share similar amino acid sequences and structure domains. Both FRS2α and FRS2β have multiple tyrosine phosphorylation sites that constitute binding sites for downstream signaling molecules for the FGF pathway. FRS2α is also phosphorylated by MAP kinase on multiple threonine residues in response to FGF and other growth factor stimulation, which relay the negative feedback control of the FGF signaling axis via the ubiquitination pathway. Germline deletion of FRS2α causes early embryonic death. Cell type specific deletion leads to congenital defects in multiple organs. Over expression of FRS2α, on the other hand, is associated with multiple types of cancer. All these clearly demonstrate the central roles of FRS2α in the FGF signaling axis.

*College of Pharmacy, Wenzhou Medical University, Wenzhou, Zhejiang, China; E-mail: wangcong0814@163.com

†Center for Cancer and Stem Cell Biology, Institute of Biosciences and Technology, Texas A&M University, 2121 W. Holcombe Blvd., Houston, Texas; USA; E-mail: fwang@ibt.tamhsc.edu

‡Corresponding authors.

1. Introduction

Interaction of a fibroblast growth factor (FGF) with its cognate receptor (FGFR) leads to changes in the receptor dimer conformation and induces trans phosphorylation between FGFRs on tyrosine residues.[1] This activates the kinase activity and creates binding sites for downstream effectors that contain a SRC homology 2 (SH2) domain or a phosphotyrosine binding (PTB) domain. Seven major tyrosine autophosphorylation sites have been identified in the intracellular domain of type 1 FGFR (FGFR1).[2] Among them, phosphorylation of tyrosine 653 and 654 is required for the removal of the autoinhibitory domain from the substrate-binding sites while phosphorylation of tyrosine 766 generates the binding site for recruitment and activation of PLCγ. Although Tyr766 is not essential for the mitogenic activity of FGFR1, it is required for the time-dependent acquisition of the proliferative response to FGFR1.[3] The function of other phosphorylation sites has not been clearly established, although there is some evidence that links these residues to the mitogenic activity of FGFRs.[4]

In general, the majority of FGF signaling has been found associated with either the MAP kinase or PI3K/AKT pathway. Instead of directly binding to the FGFR kinases, the adaptor molecules involving activation of both MAP kinase and PI3K/AKT pathways are recruited to the FGFR kinases via a scaffold protein called FGF receptor substrate 2 (FRS2). This places FRS2 front and center in control of FGFR signaling.

The FRS2 family is composed of two highly homologous members, FRS2α and FRS2β, which belong to a category of adaptor proteins that have binding sites for both upstream and downstream molecules in the signaling pathway, and physically connect downstream molecules to the upstream molecules. Depleting FRS2α abrogates the activation of the MAP kinase and PI3K/AKT pathways. Although it is not clear whether the two FRS2 isomers have independent functions, they appear functionally redundant in that expression of FRS2β in FRS2α-deficient cells restores the ability of FGFR1 to activate both MAP kinase and PI3K/AKT pathways. In addition, FRS2α is also engaged in feedback regulation of the FGF

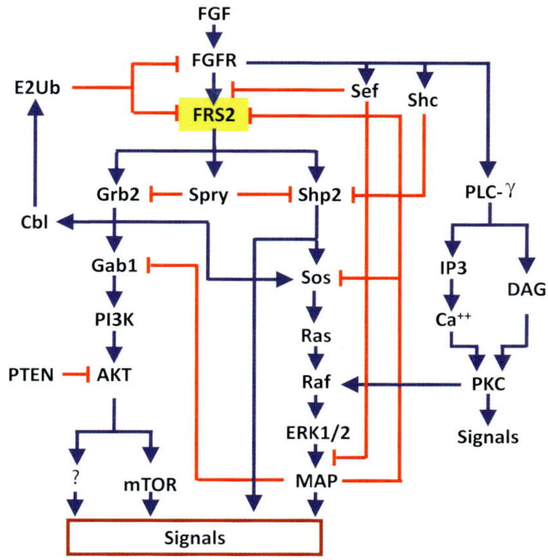

Fig. 1. Signaling pathways by FGFR tyrosine kinase. Blue lines represent positive regulation, and red lines represent negative effects.

signaling pathway. As illustrated in Fig. 1, FRS2 appears to be the key adaptor protein in the FGFR signaling cascade that mediates multiple downstream pathways of the FGFR, as well as controls the amplitude of FGFR signaling. However, whether FRS2 is also involved in signaling specificity among the different FGFR kinases or FGFR signaling in general has not been clarified.

2. Discovery and Characterization of FRS2

FRS2α was first identified as a p13^{suc1}-binding protein in PC12 pheochromocytoma cells that is rapidly phosphorylated on tyrosine residues following treatment with nerve growth factor (NGF) or FGF and designated as SNT1 (suc-associated neurotrophic factor-induced tyrosine-phosphorylated target).[5] Further studies showed that SNT1 is widely expressed and binds to an intracellular adaptor Grb2.[6] At the same time, a protein tyrosine-phosphorylated in response to FGF or NGF stimulation was purified from NIH-3T3 cells and designated as FRS2α turned out to be identical to SNT1.[7] Analysis of the full length FRS2α protein revealed the presence of a myristylation site, phosphorylated tyrosine binding (PTB) domain, and multiple tyrosine phosphorylation sites.

Overexpression of the FRS2α enhanced FGF signaling both in terms of activating various MAP kinases and promoting cell proliferation.[7] Yeast two-hybrid protein-protein interaction assay showed that FRS2α binds to the juxtamembrane segment of FGFR1, which does not contain a phosphorylated tyrosine residue.[8] Using the same strategy, the second member of the SNT family was identified and designated as SNT2 and later FRS3 or FRS2β.[8]

Both FRS2α and FRS2β are expressed in mouse embryos with FRS2α having a broader expression pattern than FRS2β. FRS2α expression can be detected as early as E7 in the developing syncytiotrophoblast, and then in the neural tube (NT) and in almost all adult and fetal tissues that have been examined. FRS2β is expressed predominantly in the ventricular layer of the developing NT and brains of murine embryos. In adult mice, FRS2β is expressed in the kidney, spleen and weakly in the brain, lung, large intestine, testis, stomach, and bone marrows.[9]

3. Protein and Genomic Structures of FRS2

FRS2α contains 508 amino acid residues and has a predicted molecular mass of 65.8 kDa.[7] FRS2β contains 492 amino acid residues with a predicted molecular weight of 54.5 kDa.[8] The two FRS2 isoforms have

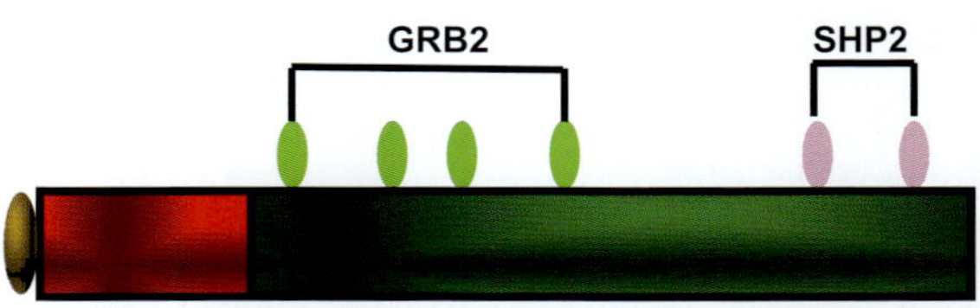

Fig. 2. Structure domains of the FRS2α. Boxes indicate the putative structure domains. Orange oval, myristylation site; red box, phosphotyrosine binding (PTB) domain; green box, binding domain for downstream signaling molecules; green ovals, GRB2-binding phosphorylation sites; purple ovals, SHP2-binding phosphorylation sites.

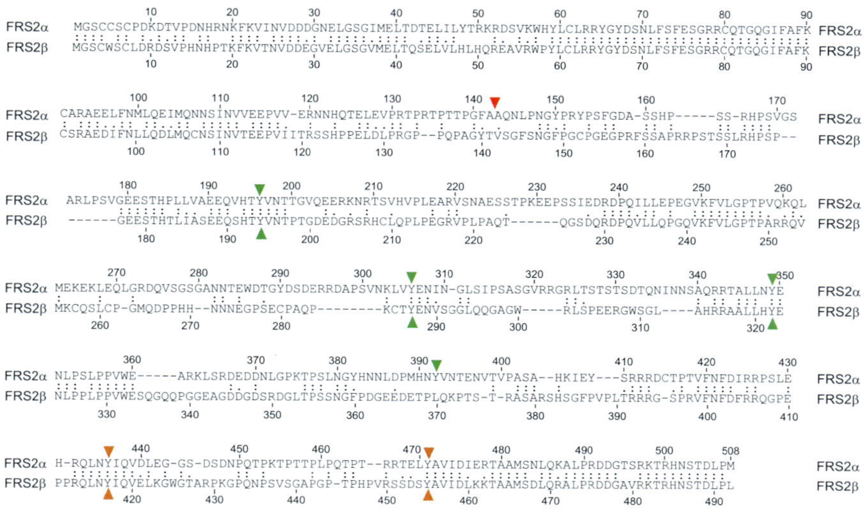

Fig. 3. Amino acid sequence of human FRS2α and FRS2β. Double dots indicate conserved amino acids: single dots, indicate semi-conserved amino acids; green triangles, GRB2-binding phosphorylation sites; orange triangles, SHp2-binding phosphorylation sites; red triangle indicates the end of the PTB domain.

a 49% sequence identity. Both FRS2α and FRS2β have an N-terminal myristylation (MGXXXS/T) site that plays a role in plasma membrane localization and a PTB domain involved in protein-protein interactions (Fig. 2). The PTB domain is highly conserved, with the similarity in amino acid sequence between FRS2α and FRS2β reaching 72% (Fig. 3). The sequence downstream of the PTB domain encodes the effector domain that includes multiple tyrosine and serine/threonine phosphorylation sites. The phosphotyrosine residues constitute binding sites for downstream signaling molecules and the phosphorylated threonine residues are involved in negative feedback controls of FGF signaling. Although the three dimensional structure of the PTB domain was defined more than a decade ago,[10–12] the structure of the C-terminal effector domain remains to be determined. FRS2α has 6 tyrosine phosphorylation sites (Tyr 196, 306, 349, 392, 436, and 473), whereas FRS2β has only 5 tyrosine phosphorylation sites (Tyr 194, 287, 322, 417, and 455). The residues surrounding each tyrosine phosphorylation site are similar. Among the 6 tyrosine

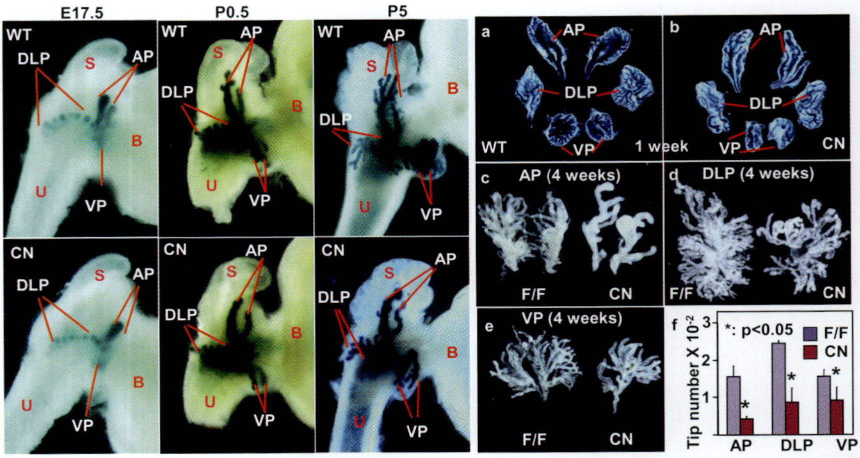

(a) (b)

Fig. 4. Disruption of FRS2α inhibits ductal branching morphogenesis in prostates. (a) The urogenital tract was dissected at the indicated times, lightly fixed, and stained with X-Gal. Stained tissues representing each prostatic lobe are indicated. (b) Prostatic lobes dissected from mice at the indicated ages were lightly fixed and stained with X-Gal (a, b). The ductal network in each lobe from 4-week-old mice was microdissected (c–e), and the average number of tip was quantified from 3 prostates and shown as mean and standard deviation (f). AP, anterior prostate; DLP, dorsolateral prostate; VP, ventral prostate; B, bladder; S, seminal vesicles; U, urethra; F/F, homozygous FRS2α^{flox} mice; CN, FRS2α^{cn} mice; WT, wildtype FRS2α.

phosphorylation sites, the first four phosphotyrosine motifs mediate recruitment of GRB2 and the two phosphotyrosine-comprising motifs at the C-terminus mediate recruitment of SHP2, which in turn helps recruiting SOS to the FGFR kinase.[13] Similarly, FRS2β has 2 SHP2-binding and 3 GRB2-binding phosphotyrosine sites. In addition, a highly conserved motif at the C-terminus also serves as a functional domain in recruitment of SOS.[13]

The genomic sequences and exon-intron organizations of the FRS2α and FRS2β alleles are highly conserved between mouse and human. Both FRS2α and FRS2β contain 5 coding exons spanning over 7 kb of genomic sequence with similar exon size and organization. Furthermore, FRS2α and FRS2β also share highly conserved non-coding sequences, suggesting their potential function as regulatory elements.[14] FRS2α encodes two transcripts with approximate sizes of 7.5 and 6.25 kb, whereas FRS2β encodes a single transcript corresponding to approximate 2.4 kb. The significance of FRS2α splice variation remains to be determined.

4. FRS2 is an Adaptor in the Receptor Tyrosine Kinase Signaling Cascade

FGF stimulation leads to phosphorylation of FRS2α, which recruits the GRB2/SOS complexes directly or indirectly through SHP2[13,15–17] and links FGFR to the MAP kinase pathway.[7,18,19] It is unknown whether FRS2α and FRS2β link to the same or different signaling pathways. Both FRS2α and FRS2β are expressed in embryos at early developmental stages[9]; ablation of FRS2α causes embryonic lethality at E7.0–E7.5,[20] whereas ablation of FRS2β did not cause detectable developmental defects (Wang, unpublished). Understanding how FRS2 is phosphorylated by each FGFR isoform will shed light on the mechanism of how the four FGFR tyrosine kinases elicit receptor-specific signals and regulate a broad spectrum of function in virtually all cell types.

Phosphorylation on tyrosine residues 196, 306, 349, and 392 of FRS2α creates binding sites for GRB2,[13,17] which binds to FRS2α through its SH2 domain and is then tyrosine phosphorylated by the FGFR kinase. GRB2 forms complexes with several docking proteins including binding to the proline-rich region of GAB1 primarily via its carboxyl-terminal SH3 domain, and to SOS1 via its amino-terminal SH3 domain. In addition, both N- and C-terminal SH3 domains of GRB2 bind to CBL1.

Interaction with GAB1 leads to the recruitment and activation of PI3K; interaction with SOS1 leads to the recruitment of the RAS-MAP kinase cascades; and interaction with CBL1 leads to ubiquitination of FRS2α and FGFR1 resulting in their degradation via the ubiquitination pathway.[21] The two tyrosine phosphorylation sites (436 and 471) in the C-terminal domain are the binding sites for protein tyrosine phosphatase SHP2, which interacts with FRS2α via its SH2 domain. Association of SHP2 and FRS2α leads to tyrosine phosphorylation of SHP2 by the FGFR kinase and activation of the ERK pathway.[20,22] Although the detailed molecular mechanism underlying how SHP2 recruits MAP kinase is not fully clear, both the catalytic activity and, to a much lesser extent, the GRB2 binding-tyrosine phosphorylation sites of SHP2 are required for maximal FGF2-induced ERK activation.[23]

SHB is a ubiquitously expressed adaptor protein that contains a C-terminal SH2 domain that interacts with multiple growth factor receptor kinases, including FGFR1. SHB also has a central PTB domain and several N-terminal proline-rich motifs that mediate its interactions with multiple cell signaling molecules. Although SHB does not directly bind to FRS2α, it interacts with phosphorylated tyrosine 766 of FGFR1 and SHP2. The interaction has been shown to be critical for FGFR1-mediated FRS2α phosphorylation and maximal FGF-mediated mitogenicity via activation of the RAS/MEK/MAPK pathway.[24]

A number of studies showed that atypical PKCs regulate the interaction between FRS2α and SHP2. The protein kinase C inhibitor bisindolylmaleimide I (Bis I) abolished the FGF2-mediated association of SHP2 with FRS2α and GAB1 for proper RAS activation.[25] An FRS2α mutant with substitutions of the 4 GRB2-binding tyrosine residues with phenylalanines still undergoes FGF-induced tyrosine phosphorylation, but recruitment of GRB2 and SOS1 are diminished, whereas SHP2 recruitment is unaffected. Reciprocally, a substitution of the two SHP2-binding tyrosine residues with phenylalanines does not affect GRB2 and SOS1 recruitment, but SHP2 recruitment is abolished.[13]

FRS2α is localized exclusively to lipid rafts *in vitro* and *in vivo*. It has been shown that upon FGF treatment, GRB2 is recruited to the rafts followed by activation of MEK1/2 by different mechanisms within lipid rafts, leading to different cellular responses during signaling events. Therefore, it is suggested that the tyrosine phosphorylation and serine-threonine phosphorylation of FRS2α within lipid rafts affect the response of cells to FGF signaling and that the compartmentalized signaling within the rafts may provide a level of specificity for growth factor signaling.[26] Flotillin-1 binds to the PTB of FRS2α and competes for the binding with the FGFR. CBL-associated protein/ponsin (CAP) directly interacts with FRS2α by means of its sorbin homology (SoHo) and the third SH3 domains. Flotillin-1 is necessary for the signaling induced recruitment of FRS2 into lipid rafts. Since the binding domains for these three proteins overlap, it has been suggested that flotillin-1 and CAP compete for the binding to FRS2α and regulating FGF signaling intensity.[27]

CKS1 is a mammalian homologue of SUC1, a yeast cyclin-dependent kinase-binding protein that binds to and has been used to capture FRS2α. CKS1 is required for $p27^{Kip1}$ to interact with SKP-2 (S-phase kinase-associated protein2) ubiquitin ligase complex and promotes $p27^{Kip1}$ degradation, a key step for cells to enter the S phase of the cell cycle. CKS1 constitutively associated with unphosphorylated FRS2α. FGF-dependent activation of FGFR tyrosine kinases induces FRS2α phosphorylation, leading to release of CKS1 from FRS2α. The released CKS1, in turn, promotes degradation of $p27^{Kip1}$ in the cells. Since degradation of $p27^{Kip1}$ is a key regulatory step in activation of the cyclin E/A-CDK complex during the G1/S transition of the cell cycle, the results reveal a mechanism by which FRS2α-mediated signals directly regulate cell cycle progression.[28]

RND1 directly associates with FRS2α and FRS2β at the COOH-terminal region including tyrosine residues essential for the interaction with SHP2. Similarly, RND1 constitutively binds to FRS2α at the same site as SHP2. When FRS2α is phosphorylated by FGFR1 kinase, it recruits SHP2, and releases RND1. The liberated RND1 then inhibits RHOA activity. These results suggest that the activity of RND1 is also regulated by FGFR through FRS2α.[29]

FGF activates mTOR via the FRS2α-mediated PI3K/AKT pathway, and suppresses autophagy activity in mouse embryonic fibroblast via the PI3K/AKT pathway.[30] In addition, FRS2α is required for FGFR1 to complex with and activate mTOR in vascular smooth muscle cells required for dedifferentiation of the cells during atherosclerosis and restenosis after vascular injury.[31] CRK binds to FRS2α at tyrosine 436 to form the FRS2α-CRK complex, which is required for NGF to activate RAP1 in PC12 cells.[32] FRS2α also has been reported to interact with the Giα protein, the inhibitory alpha subunit of G protein trimers that regulates GTPase activity. Interestingly, Giα has been shown required for FRS2α to associate with GAB1 as well as GAB1 with GRB2. Ablation of Giα impairs the formation of the FRS2α-GRB2-GAB1 complex and diminishes activation of the AKT and mTORC1 pathways by the FGF.[33]

Although FRS2α and FRS2β share a high homology in their amino acid sequences, the two members have distinct expression patterns. Furthermore, FRS2α has 6 tyrosine phosphorylation sites and FRS2β has 5 tyrosine phosphorylation sites. Ablation of FRS2α leads to early embryonic lethality,[20] whereas ablation of FRS2β does not lead to detectable defects in all organs examined (Wang, unpublished data). This indicates that FRS2α and FRS2β have distinct functions in embryonic development. However, expression of FRS2β can compensate the loss of FRS2α for activation of MAP kinase in mouse embryonic fibroblasts.[34] It has been reported that FRS2β binds microtubules comparable to the microtubule-associated protein, MAP2, while FRS2α does not.[35] Several reports also demonstrate the role of FRS2β in MAPK-mediated negative feedback controls of receptor tyrosine kinase signaling. This is discussed in more detail in a subsequent section.

The neurotrophins are a family of closely related growth factors that regulate proliferation and differentiation of neuronal cells via activating TRK receptors. Treating primary cortical neurons with neurotrophins stimulates the tyrosine phosphorylation of FRS2α and the subsequent recruitment of SHP2, and GRB2. Binding of GRB2 recruits GAB1 and GAB2 to FRS2α, providing an alternative route to activate PI3 kinase and SHP2. Overexpressing FRS2α, but not FRS2β, promotes neurite outgrowth and branching in cortical neurons relative to controls.[36] Together, these data suggest that FRS2α and FRS2β have both overlapping and distinct functions in cell signaling networks.

5. Interaction of FRS2 with FGFR Kinases

FRS2α has a central role in FGF signaling pathways. FRS2α binds to the intracellular juxtamembrane domain of the FGFR kinase via its

PTB domain.[8,37] In general, PTB domains play a similar role to SH2 or PH domains that recognize a motif in the target proteins and share a similar global structure.[38–40] However, unlike SH2 domains that only recognize phosphotyrosine-containing motifs, the PTB domain recognizes a consensus motif NPXY where the tyrosine residues may or may not be phosphorylated. Interestingly, the PTB-binding site on FGFR1 does not have the conserved NPXY motif, indicating that FRS2α binds to FGFR kinases differently than the classical PTB domain binding to PTB-recognition domains. Crystal and NMR structures and mutational analyses reveal that the PTB domain of FRS2α utilizes a distinct set of amino acid residues to interact with FGFRs and TRK.[37] Therefore, it is suggested that FRS2α may relay receptor-specific signals and serve as a molecular switch between distinct signals from FGFR and TRK.

The interaction between the PTB domain and the intracellular juxtamembrane domain has been demonstrated in several ways that included the yeast two hybrid assay system, direct pull-down assay, and NMR analyses.[8,10,41] The alternatively spliced VT (valine and threonine) domain in the intracellular segment has been reported to be required for the interaction between the PTB domain and the juxtamembrane domain of FGFR kinases.[41] The VT^+ and VT^- isoforms of FGFR1 are differently expressed in xenopus embryos and elicit different function to mediate mesoderm formation.[42] Deleting the two amino acids in FGFR1 disrupts FGFR1 signaling and leads to embryonic development defects.[43] However, *in vitro* analyses show that both VT^+ and VT^- isoforms also binds to FGFR1 albeit with a lower efficiency than the VT^+ isoform.[44] Although the PTB domain binds to the FGFR1 kinase constitutively, full-length FRS2α binding to FGFR1 is enhanced by activation of the receptor kinases, suggesting that the C-terminal sequence downstream of the PTB domain inhibits PTB-FGFR1 binding. Inactivation of the FGFR1 kinase and substitution of tyrosine phosphorylation sites of FGFR1, but not FRS2α, reduces binding of FGFR1 with FRS2α. The results suggest that although the tyrosine phosphorylation of FGFR1 does not constitute the binding sites for FRS2α, it enhances the interaction between FRS2α and FGFR1 kinase.[44]

The quantity and quality of phosphorylation of FRS2α is both FGFR isotype- and cell type-specific.[45] In prostate tumor epithelial cells, the resident FGFR2 in epithelial cells weakly phosphorylates FRS2α and the resulting phosphorylated FRS2α has an apparent molecular weight of 70-kDa as indicated by SDS-PAGE. However, ectopic FGFR1 in the epithelial cells strongly phosphorylates FRS2α, which yields a phosphorylated FRS2α with an apparent molecular weight of 85-kDa on the SDS-PAGE. In addition, the phosphorylation of FRS2α by FGFR1 kinase, both of which are membrane-anchored proteins, only occurs in intact cells. The differential phosphorylation of FRS2α is likely a determinant of the specificity of signaling by FGFR. Therefore, in addition to differences in the structure of intracellular kinase domains among FGFR isotype subunits within oligomeric heparan sulfate-FGFR complexes, cell membrane and cytoskeletal context likely also determine FGFR isotype- and cell type-specific organization and conformational relationships between FGFR kinases and external substrates. The precision of conformational relationships among kinase subunits and proximal substrates rather than simple concentration and proximity are the rate-limiting determinant of basal activity and amplitude of FGFR signaling as well as FGFR isotype- and cell type-specificity. Understanding how FRS2α contributes to the receptor-specific signaling of FGFR tyrosine kinases in cells will provide a molecular mechanism by which FGFs elicit specific signaling that controls a broad spectrum of functions in diverse cell types and tissues.

6. Interaction of FRS2 with other Growth Factor Receptors

In addition to FGFR receptor tyrosine kinases, FRS2α binds to several other RTKs via its PTB domain either in a phosphotyrosine dependent and independent manner. These include neurotrophin receptors (TRKA, TRKB, and TRKC), ephrin receptor (EphA4), and anaplastic lymphoma kinase (ALK). However, FRS2α has been shown only to be phosphorylated by FGFR, TRKs, and IRs. FRS2α becomes tyrosine phosphorylated by EGFR kinases only in A431 and NIH3T3 cells where the EGFR is overexpressed,[46] but not in other cells expressing EGFR at modest levels.

FRS2α interacts with TRKA, becomes tyrosine-phosphorylated, and then serves as a scaffold for organization of a number of signaling molecules including GRB2, SHP2 CRK, C3G, RAP1, and B-RAF upon NGF stimulation.[7,22,32,47,48] It mediates differentiation signals of TRK in PC12 cells.[5,7] Although the PTB domain is also responsible for FRS2α to bind to TRK, as mentioned earlier it utilizes different sets of amino acid residues for interacting with FGFR and TRK in a mutually exclusive manner.[10,12,49] In fact, it has been suggested that FRS2α functions as a molecular switch between FGFR and NGFR. NGF-induced tyrosine phosphorylation of FRS2α is diminished in cells overexpressing a kinase-inactive mutant of FGFR1, suggesting that FGFR1 regulates NGF signaling receptors by sequestering a common key element which both receptors utilize for transmitting their signals.[50] FRS2α binds to and is phosphorylated by TRKB upon BDNF treatment, which then recruits GRB2 and the RAS/MAPK pathway for activation by TRKB.[51,52] FRS2β also binds to all TRK isoforms and becomes tyrosine phosphorylated in response to NGF, BDNF, and NT-3.[53] TRKB-mediated tyrosine phosphorylation of FRS2β results in binding of FRS2β to SHP2.[54] However, overexpression of FRS2β in PC12 cells neither increases NGF-induced neuritogenesis nor activates MAPK/AKT in PC12 cells. To date, the function of FRS2β in the TRK signaling cascade still remains elusive.

RET is the receptor for glial cell line-derived neurotrophic factor (GNDF). The FRS2α PTB domain binds to RET at the phosphorylated tyrosine 1062 that is the same site used by the SHC adaptor protein.[15,55] Although both SHC and FRS2α can be phosphorylated by RET, SHC binds to GRB2 and GAB1 proteins and mediates both MAPK and PI3K/AKT pathways, whereas FRS2α binds to GRB2 and only mediates the MAPK pathway, but not the PI3K/AKT pathway for RET.[55] Engagement of RET with FRS2α recruits RET to lipid rafts. In addition, FRS2α also recruits RET to focal adhesion complexes and leads to activation of SRC family kinases and focal adhesion kinase (FAK) required for cell migration. Together, the results suggest that

FRS2α concentrates RET in specific membrane foci and contributes to the signaling specificity of RET.[56,57]

In vitro experiments show that VEGF induces FRS2α phosphorylation, and forms a complex with NCK, PAK, GRB2, PKCl, and CRK, although direct interaction of VEGFR and FRS2α was not shown.[58] Expression of FRS2α is required for VEGF-A and VEGF-C to induce VEGFR autophosphorylation and affects VEGF-A and VEGF-C dependent activation of ERK signaling, as well as migration and proliferation of blood and lymphatic endothelial cells. Endothelial-specific deletion of FRS2α results in the profound impairment of postnatal vascular development and adult angiogenesis, lymphangiogenesis, and arteriogenesis, suggesting that FRS2α is a component of the VEGFR signaling pathway.[59]

It has been shown that FRS2α associates with insulin receptor (IR) and can be phosphorylated by the IR. Furthermore, insulin stimulates tyrosine phosphorylation of FRS2α and its subsequent association with SHP2. The results suggest that FRS2α can participate in insulin signaling by recruiting SHP2 and, hence, function as a docking molecule similar to insulin receptor substrate proteins.[60]

FRS2α can be pulled down by PDGFR in a complex that requires FGFR1. Both the extracellular and the intracellular domains of FGFR1 are required for association with PDGFR, whereas the cytoplasmic domain of FGFR1 is required for FRS2α association with the FGFR1-PDGFR complex. Knockdown of FRS2α in vascular smooth muscle cells (VSMC) by RNA interference inhibited PDGF-BB-mediated down-regulation of smooth muscle actin (SMA) and SM22 without affecting PDGF-BB mediated cell proliferation and ERK activation. The data support the notion that PDGFR down-regulates SMA and SM22 through formation of a complex that requires FGFR1 and FRS2α providing a novel insight into mechanisms of the phenotypic plasticity of VSMC.[61]

7. FRS2 in the Feedback Control of Tyrosine Kinase Signaling

In addition to tyrosine phosphorylation that activates FRS2α for recruiting downstream signaling molecules to receptor tyrosine kinases, emerging evidence shows that FRS2α is also phosphorylated by MAP kinase on multiple threonine residues in response to FGF stimulation, insulin, EGF, and PDGF. The phosphorylation of threonine by MAP kinases creates binding sites for CBL1, an E3 ubiquitination ligase and contributes to termination or down regulation of signal from these growth factor receptors. Prevention of threonine phosphorylation of FRS2α results in its constitutive tyrosine phosphorylation in unstimulated cells and enhances tyrosine phosphorylation of FRS2α, MAPK activation, and cell migration and proliferation in FGF stimulated cells.[46,62] Unlike EGF and PDGF receptors that bind to CBL directly, FGFR does not directly bind CBL. Instead, GRB2 binds to tyrosine-phosphorylated FRS2α and forms a ternary complex with CBL via its SH3 domains, which results in ubiquitination of FGFR and FRS2α in response to FGF stimulation.[21]

FGF-induced tyrosine phosphorylation of FRS2α can be attenuated by co-stimulation with EGF in PC12 cells. The attenuation can be reversed by inhibition of MEK activity. The FRS2α mutant that lacks the MAPK binding and phosphorylation sites cannot be inhibited by EGF treatment and are more active than wildtype FRS2α in response to FGF treatment. These results show that phosphorylated threonine of FRS2α by MAPK functions as a molecular switch that negatively regulates FGF signal transduction in response to either FGF or other growth factor stimulation.[63]

UNC-51-like kinase 2 (ULK2), a member of a conserved subfamily of ubiquitously expressed serine/threonine-specific protein kinases, interacts with the PTB domain of FRS2α/β and negatively regulates tyrosine phosphorylation of signaling proteins downstream of FGFR1.[64] Activation of FGFR leads to FRS2α-dependent activation of the SAC kinase, which in turn phosphorylates the signaling attenuator human Sprouty 2 (SPRY2) on residue Y55. The phosphorylation enhances formation of the SRC-SPRY2 complex and is required for SPRY2 to inhibit activation of the ERK pathway via inhibiting GRB2 binding to FRS2α. This results in activation of the SPRY2-mediated signal attenuation pathways.[65]

Although FRS2β does not become tyrosine phosphorylated in response to EGF treatment, it forms a complex with ERK2 and formation of the complex is enhanced by EGF stimulation. FRS2β downregulates ERK2 phosphorylation and suppresses and delays ERK2 nuclear accumulation, which occurs following the EGF stimulation.[66] FRS2β constitutively binds to EGFR and ErbB2 through the PTB domain, as well as the trafficking of CD2AP and CIN85 that target lysosomes and CBL ubiquitin ligase. The complex of FRS2β-CIN85/CD2AP-CBL functions as a feedback control downregulating ErbB2 protein.[67] Expression of FRS2β attenuates FGFR signaling including receptor autophosphorylation, suppressed EGF-induced cell transformation and proliferation. The attenuation can be alleviated by treatment of cells with MEK inhibitor U0126. The results suggest a role of FRS2β in negative controls on EGFR and ErbB2 activities.[68]

PEA15 is a widely expressed and highly conserved protein that comprises an N-terminal death effector domain and a largely unstructured C-terminal tail. Aberrant PEA15 expression has been implicated in numerous pathologies, including cancers from various tissue origins. PEA15 binds directly to ERK1/2 and limits ERK1/2 entry to the nucleus by blocking nuclear import and promoting nuclear export, and thus, sustains ERK1/2 activation. In addition, PEA15 binding prevents ERK1/2 membrane recruitment and threonine phosphorylation of FRS2α. Threonine phosphorylation of FRS2α proceeds the CBL1-dependent feedback downregulation of FRS2α. Thus, this reduced threonine phosphorylation leads to increased FGF-induced tyrosine phosphorylation of FRS2α and enhances downstream signaling. Deletion of FRS2α blocks the capacity of PEA15 to activate the MAP kinase pathway. Therefore, PEA15 dominantly sustains ERK1/2 activation by inhibiting ERK1/2-dependent threonine phosphorylation of FRS2α.[69]

8. FRS2α in Embryonic Development

FRS2 is a major mediator of FGF signaling by providing a link between FGFRs and downstream intracellular signaling pathways via its 6 tyrosine phosphorylation sites. Among them, four are the binding sites for GRB2 and two for SHP2, which in turn recruit proteins initiating the RAS-ERK and PI3K-AKT signaling pathways. Germline ablation of FRS2α leads to embryonic lethality at E7.0–7.5. Knocking in of an FRS2α mutant lacking the SHP2 binding sites affects viability of mouse embryos at E9.5, although some mutant embryos can survive until E18.5.[70] Deletion of the four GRB2-binding sites, however, does not cause detectable developmental defects in major organs, except abnormal eyelids at birth. The SHP2-binding site mutants exhibit a variety of developmental defects, such as eye, branchial arch, limb, brain, nerve, thyroid, ultimobranchial body, parathyroid, and thymus, and heart.[70–72] The defect in brain development appears to be due to, at least in part, reduced neural stem/progenitor cells (NSPC), suggesting that FRS2α plays critical roles in the maintenance of intermediate progenitor cells and in neurogenesis in the cerebral cortex. However, mutant NSPC appears to be able to self-renew, demonstrating that SHP2-binding sites on FRS2α are important for NSPC proliferation, but are dispensable for self-renewing capacity of NSPC after FGF2 stimulation.[71] The carotid body is a small cluster of chemoreceptors and supporting cells located near the fork (bifurcation) of the carotid artery, which is a chemosensory organ that is sensitive to arterial hypoxia, hypercapnia, and acidity. It plays a role in the regulation of the respiratory system. There are two types of cells in the carotid body, glomus cells and sustentacular cells. Replacing the wildtype FRS2α with the SHP2 binding site mutant abrogates carotid body development.[73] These results suggest that the FRS2α-SHP2-MAPK pathway is required for the carotid body development.

Since germline disruption of FRS2α is early embryonic lethal, a loxP-flank FRS2α allele has been generated for tissue-specific ablation of FRS2α in mice, in which the sequence encoding the structure downstream the PTB domain is flanked by a pair of loxP elements.[74] The insertion of loxP elements does not lead to any detectable defects. However, a cross with mice carrying tissue-specifically expressed Cre recombinase to delete the loxP flanked sequence leads to severe defects in a variety of organs.

In mice development of the lens starts at E10.5 when the invaginating lens ectoderm detaches from the anterior head epithelium to give rise to the lens vesicle, followed by differentiation of the posterior lens vesicle. The subsequent growth of the lens requires proliferation and migration of the anterior lens epithelial cells. FGF signaling has been shown to be necessary and sufficient for inducing lens fiber development. Using an array of genetically engineered mouse models combined with biochemical analyses, Li, *et al.*, show that double depletion of FRS2α and SHP2 results in a more severe FGF signaling defect than either single mutant. This synergy between FRS2α and SHP2 is not a result of overlapping or compensatory functions of the two factors. Instead, FRS2α acts directly upstream to SHP2 in a linear cascade to mediate FGF signaling. Both constitutively activated SHP2 and KRAS signaling can ameliorate lens defects in FRS2α mutants. Since, both GAB1 and GAB2, the commonly believed downstream target of FRS2α, is not required for FGF signaling to regulate lens development, these results establish the FRS2α –SHP2 complex as the key mediator of FGF signaling in lens development.[75]

Inner ear mechanosensory hair cells transduce sound and balance information. FGFR1 signaling through FRS2α and FRS2β is necessary for prosensory formation. In the absence of FGFR1-FRS2 signaling, the prosensory domain becomes post-mitotic. However, the expression of prosensory markers is impaired, resulting in fewer sensory precursors and less HC. In addition, the activation of the MAPK pathway and the expression of SOX2 are compromised. These results show that FRS2 is essential for mediating FGF signals necessary for the maintenance of sensory progenitors and commits precursors to sensory cell differentiation in the mammalian cochlea.[76]

The cardiac outflow tract (OFT) is a developmentally complex structure derived from multiple lineages of cells and is often defective in human congenital anomalies. FRS2α is required for mediating crucial aspects of FGF-dependent OFT development in mouse. Ablation of FRS2α in mesodermal OFT progenitor cells that originate in the second heart field (SHF) affects their expansion into the OFT myocardium, resulting in OFT misalignment and hypoplasia (Fig. 5). Moreover, ablation of FRS2α also causes defective endothelial-to-mesenchymal transition and neural crest cell recruitment into the OFT cushions, resulting in OFT septation defects.[77] In addition, FRS2α is required for FGF signaling in the OFT

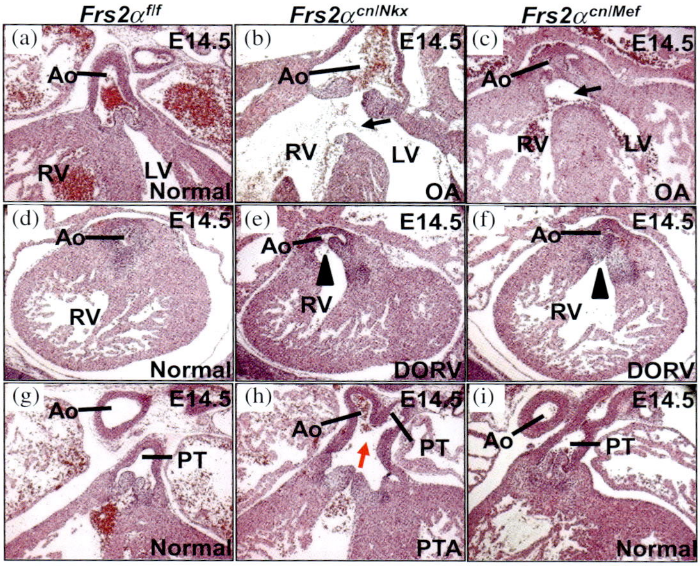

Fig. 5. Outflow tract (OFT) alignment and septation defects in Frs2α^{cn} mutants. H&E staining of E14.5 embryonic heart sections demonstrates overriding aorta (a-c) and double outlet right ventricle (d-f) defects in FRS2α$^{cn/Nkx}$ and FRS2α$^{cn/Mef}$ mutants, and persistent truncus arteriosus defects in FRS2α$^{cn/Nkx}$ mutant (g-i). Ao, aorta; DORV, double outlet right ventricle; LV, left ventricle; OA, overriding aorta; RV, right ventricle; PT, pulmonary trunk; PTA, persistent truncus arteriosus. Black arrows denote OA-associated ventricular septal defects; red arrow denotes the persistent truncus arteriosus; arrowheads denote double outlet right ventricle.

myocardium to upregulate Bmp4 expression, which then enhances smooth muscle differentiation of neural crest cells (NCCs) in the OFT cushion, as well as to promote OFT myocardial cell invasion to the cushion. Disrupting FRS2α in the SHF interrupts cushion remodeling with reduced NCCs differentiation into smooth muscle and less cardiomyocyte invasion, resulting in malformed OFT valves (Fig. 6). The results reveal that FRS2α -mediated signals crosstalk with BMP signaling and regulate formation of OFT valve primordia by controlling smooth muscle differentiation of NCCs in the cushion.[78] Furthermore, FRS2α -mediated signals in the SHF are required for preventing heart progenitor cells from undergoing premature differentiation by inhibition of autophagy. Disruption of

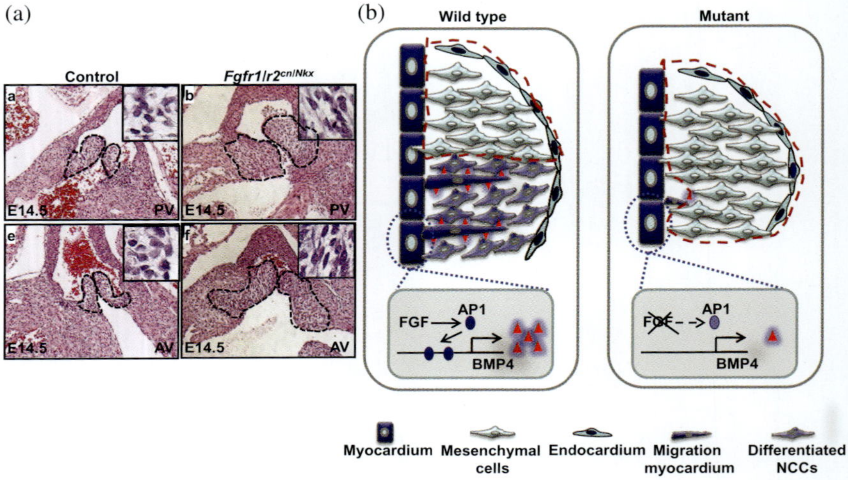

Fig. 6. Disruption of the FGF signaling axis leads to OFT valve hyperplasia. (a) Transverse sections of E14.5 embryos were H&E stained to demonstrate enlarged OFT valves. Pulmonary and aortic valves are highlighted with dotted lines. Inserts are enlarged pictures of the valve. (b) Schematic of FGF signaling in OFT valve formation. Defective cushion NCC differentiation fails to define valve primordia and leaves excessive cells within valve primordia, which results in a large valve. Valve primordia are outlined with red lines. PV, pulmonary valve; AV, aortic valve.

FGF signaling leads to premature differentiation of cardiac progenitor cells in mice. (Fig. 7). Interestingly, FGF signaling promotes mesoderm differentiation in embryonic stem cells in embryoid body cultures. The results indicate that the FGF/FRS2α -mediated signals elicit stage-specific roles in regulating cardiomyocytes.[79]

Maintenance of normal endothelial function is critical to various aspects of blood vessel function, but its regulation is poorly understood. Chen, *et al.*, showed that FRS2α is required for FGF signaling in the endothelium to upregulate TGF-β, which prevents endothelial-to-mesenchymal transition (Endo-MT).[80] The results establish FRS2α -mediated FGF signaling as a critical factor in maintenance of endothelial homeostasis.

In normal prostate epithelial cells, FGFR2 is the resident isoform playing critical roles in prostate development and maintaining tissue homeostasis and function of the prostate.[81–83] FRS2α is uniformly

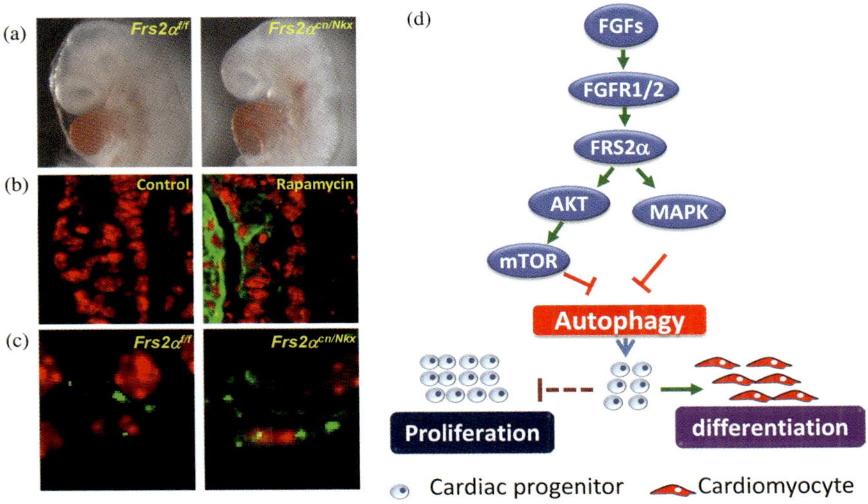

Fig. 7 . FRS2α-mediated signals suppress premature differentiation of second heart field progenitor cells through regulating autophagy activity. (a) Whole-mount immunostaining of E9.0 embryos showing premature expression of myosin in the second heart field (SHF) progenitor cells of FRS2α conditional null embryos. (b) Immunostaining of E8.5 embryos showing premature differentiation of SHF cells in rapamycin treated embryos. (c) Immunostaining showing that ablation of FRS2α increases autophagy the outflow tract myocardium. (d) Schematic shows the FGF signaling axis inhibits premature differentiation of cardiac progenitor cells by suppressing autophagy.

expressed in epithelial cells of developing prostates, whereas it is expressed only in basal cells of the mature prostate epithelium.[84] Ablation of FGFR2 in the prostate epithelium impairs prostate organogenesis, compromises branching morphogenesis, and abrogates the acquisition of the androgen responsiveness of the prostate.[81] Ablation of FRS2α also impairs prostatic ductal branching morphogenesis, and compromises cell proliferation (Fig. 4). Unlike the FGFR2 ablation, disrupting FRS2α has no effect on the response of the prostate to androgens.[84] Thus, it appears that FRS2α in prostate epithelial cells largely mediates signals related to mitogenicity, but not the FGFR2-elicited signals involved in tissue homeostasis. The details of the molecular mechanism underlying the differences remain unknown. Understanding how FGFR2 elicits its mitogenic and homeostasis-maintaining signals via FRS2α-dependent and independent pathways will shine new light on how normally resident FGFR signals regulate normal cellular function and aberrant FGF signaling contributes to cancer initiation and progression.

The mammalian epididymis is a long convoluted duct that develops from the anterior part of the Wolffian duct. There are two different activity levels of the components of the ERK pathway in the epididymal epithelium during development: a basal level in most regions and a higher level in the differentiated initial segment. FRS2α is required for regulating the basal activity level of ERK and is important for cell survival during the undifferentiated period of epididymal development. However, FRS2α is not involved in regulating the high activity level of ERK pathway and is not needed for cell survival past the period of differentiation.[85] Therefore, the results highlight the effects of the FGF signaling on promoting cell proliferation during epididymis development.

FRS2α is also expressed and regulates development in lower organisms. In Xenopus, the FRS2α like protein xFRS2 has an expression pattern that is quite similar to FGFR1 during Xenopus development and plays a critical role in the appropriate formation of mesoderm-derived tissue during embryogenesis.[86,87] It is tyrosine phosphorylated in early embryos, and overexpression of an unphosphorylatable form of xFRS2 interferes with FGF-dependent mesoderm formation.[88] The SRC family kinase Laloo, which was shown to function in FGF signaling during early Xenopus development, binds to xFRS2 and promotes tyrosine phosphorylation of xFRS2. Moreover, xFRS2 and Laloo bind to Xenopus FGFR1. Depleting xFRS2 expression or overexpression of a dominant negative mutant of xFRS2 leads to severe malformation of trunk and posterior structures, as well as disrupts muscle and notochord formation, and inhibits FGFR-induced MAP kinase activation. This indicates that xFRS2 is a critical mediator of FGF signaling and is required for early Xenopus development.[86,87] In addition, xFRS2 mediates germinal vesicle breakdown induced by an activated FGFR1 in Xenopus oocytes.[89] These results suggest that xFRS2 plays an important role in FGF signaling in cooperation with Laloo during development.

Although an FRS2α like gene is also expressed and functions upstream of the RAS/MAPK pathway required for oocyte maturation in *C. elegans*, FGFR does not need FRS2α to recruit the GRB2/SOS/RAS cassette for activating the MAP kinase pathway. Instead, GRB2 directly associates with FGFR at the C-terminal motif and regulates a number of processes, including sex myoblast (SM) migration guidance and fluid homeostasis.[90]

9. Aberrant FRS2α-mediated Signaling and Cancer

As the central adaptor for the FGF signaling pathway that contributes to initiation and progression of cancers from various tissue origins, emerging evidence shows that FRS2α is overexpressed in multiple cancers. The FRS2α coding sequence is located at chromosome 12q15, which is frequently amplified in liposarcomas,[91] embryonal rhabdomyosarcoma cell line RMS-YM,[92] adrenocortical cancer,[93] and glioma.[94] In addition, high-resolution genomic mapping reveals consistent amplification of the FRS2α gene in well-differentiated and dedifferentiated liposarcoma.[95] Analysis of the Cancer Genome Atlas (TCGA) datasets spanning 16 cancer subtypes has identified 42 putative cancer driver genes linked to diverse oncogenic processes. Among them, FGFR and its adapter protein, FRS2α, are frequently co-amplified.[96] As gene amplification represents one of the molecular mechanisms of oncogene overexpression in many types of tumors, the findings suggest the oncogenic roles of FRS2α can be a potential cancer drug target.

FRS2α is not expressed in resting mature prostate luminal epithelium.[84] However, it is strongly expressed in luminal epithelial cells of regenerating prostates and prostate tumors. Ablation of FRS2α inhibits prostatic tumorigenesis and progression of PCa and shortens the lifespan of the TRAMP mice (Fig. 8), which is a prostate tumor model induced by expression of oncogenic viral proteins in prostate epithelial cells. Although FRS2α and FRS2β are not overexpressed in human prostate cancer cells (LNCaP, DU145, and PC3) or immortalized human prostate epithelial cells (PNT1A and PNT2) derived from para-cancerous tissues, targeted dual

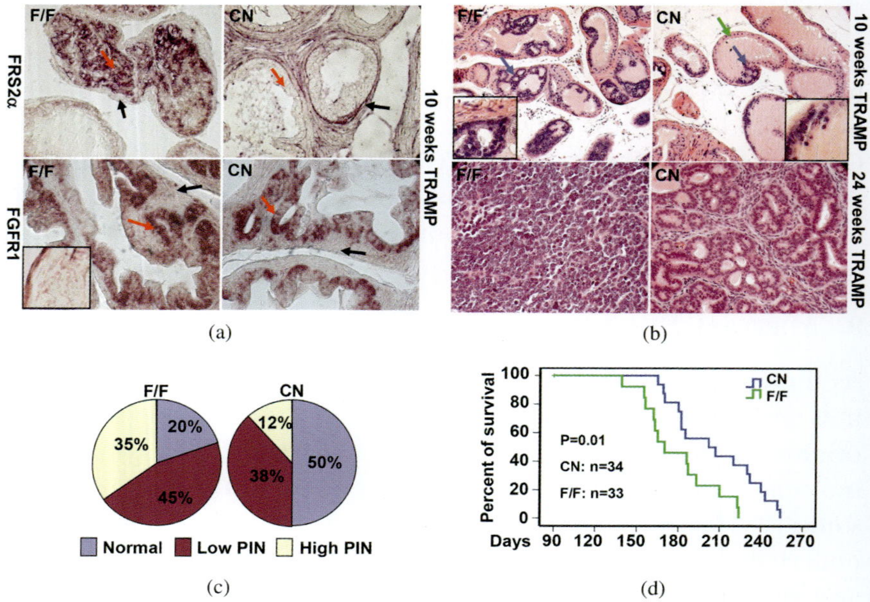

Fig. 8. Ablation of FRS2α in the prostatic epithelium inhibited tumorigenesis in the TRAMP prostatic tumor model. (a) Expression of FRS2α and *Fgfr1* at the mRNA level was assessed in TRAMP prostates by *in situ* hybridization. Red arrows indicate epithelial cells, and black arrows indicate stromal cells. Insert, wildtype control showing no expression of *Fgfr1* in the epithelial compartment. (b) Prostate tissue sections were prepared from TRAMP mice with or without FRS2α^{CN} alleles at the indicated ages and stained with H&E. (c) The percentage of areas occupied by the prostatic intraepithelial neoplasia (PIN) foci in prostates of 10-week-old TRAMP mice. The data are mean and standard deviation from 5 prostates. (d) Mortality of TRAMP mice with the indicated FRS2α alleles was determined from daily observation over 250 days. The percentage of mice that survived to the respective age is shown. F/F, homozygous TRAMP-FRS2α^{flox} mice; CN, TRAMP-FRS2α^{cn} mice.

inhibition of FRS2α and FRS2β selectively and adversely affect malignant, but not benign prostate tumor cells.[97] Hyperactivated FRS2α-mediated signaling in prostate cancer cells promotes tumor angiogenesis and predicts poor clinical outcome of patients.[98] Micro RNA miR-206 is downregulated particularly in gastric cancer with lymphatic metastasis, local invasion, and advanced TNM staging. It has been shown to have anti-metastatic effects that likely occur by downregulating signaling molecules, including FRS2α, that contribute to cancer metastasis.[99] Inhibiting FGFR signaling using shRNA and dovitinib achieved significant anti-cancer cancer effects in pancreatic cancer. The effect was more pronounced in pancreatic cancers overexpressing FGFR2IIIb that may be dependent on aberrant stimulation by stromal-derived FGF ligands.[100] Expression of the PTB domain of FRS2α inhibits activation of MAP kinase and PI3K/AKT pathways and suppresses the antiestrogen-resistant growth induced by FGF-1 in human breast carcinoma cells.[101]

The thyroid TRK oncogenes are generated by chromosomal rearrangements juxtaposing the neurotrophic receptor TRK1 tyrosine kinase domain to foreign activating sequences. The rearrangement leads to a constitutive active TRK kinase. Both FRS2α and FRS2β can be recruited and activated by TRKs at the same tyrosine residue that also interacts with SHC. Both FRS2 members and SHC are involved in coupling the receptor tyrosine kinase to the MAPK pathway by recruiting Grb2/SOS. The mutation of the docking site for both SHC and FRS2 abrogates biological activity of the oncogene.[102,103] The nucleophosmin-anaplastic lymphoma kinase (ALK) oncoprotein is a chimeric protein comprised of parts of the nuclear protein NPM and ALK receptor protein-tyrosine kinase that resulted from genomic rearrangement. NPM-ALK interacts with FRS2α and FRS2β at three sites, Y156, Y567, and a 19-amino-acid sequence motif of NPM-ALK. The three FRS2-binding sites of NPM-ALK are essential for its transforming activity.[104] Together, these data unravel the oncogenic roles of aberrant activated FRS2α pathways.

Despite accumulating evidence showing oncogenic roles of aberrant FRS2α-mediated signals, there are also reports showing that resident FRS2α-mediated signals maintain tissue homeostasis and are involved in feedback control of receptor tyrosine kinase activity thereby having a tumor suppressor function. Deletion of FRS2α from the ureteric epithelium causes renal hypoplasia. Mice with a conditional deletion of FRS2α in the ureteric epithelium (designated as FRS2UB/) developed mild renal hypoplasia characterized by decreased ureteric branching morphogenesis, but maintained normal overall branching architecture and had normal mesenchymal stromal development. Reduced nephron endowment in postnatal mutant mice was observed, corresponding with the reduction in branching morphogenesis. Ret and Wnt11 mRNA levels were lower in mutants as opposed to controls. Taken together, FRS2α may transmit signals through other receptor kinases present in ureteric epithelium. Finally, the renal hypoplasia observed in FRS2UB/ mice is likely secondary to decreased Ret and Wnt11 expression.[105] The expression level of FRS2β is downregulated in cancer.[68] Overexpression of FRS2β inhibits anchorage-independent cell growth. Knockdown of both CIN85 and CD2AP or CBL, or treatment with lysosomal degradation inhibitors diminishes FRS2β downregulation of ErbB2. In addition, knockdown of endogenous FRS2β causes upregulation of ErbB2 in primary neural cells. On the other hand, human breast cancer tissues that overexpress ErbB2 only express FRS2β at low levels.

FRS2β inhibits epidermal growth factor-receptor (EGFR) tyrosine kinase without being phosphorylated at tyrosine residues after EGF stimulation by binding to ERK and inhibiting heterodimer formation between EGFR and ErbB2. In response to EGF, ERK translocated to the plasma membrane in cells expressing FRS2β but not an FRS2β mutant in which four arginine residues in the D domains were replaced with alanines. This suggests that FRS2β serves as a plasma membrane anchor for activated ERK. In this study, FRS2β inhibits anchorage-independent cell growth induced by oncogenic ErbB2, another member of the EGFR family. Finally, a low mRNA expression level of FRS2β correlated with poor prognosis in a cohort of 60 non-small cell lung cancer patients. Therefore, FRS2β acts as a feedback inhibitor of EGFR family members and is a tumor suppressor.[106] Understanding how resident FRS2-mediated signals maintain tissue homeostasis and aberrant FRS2α-mediated signals promote cancer initiation and progression not only will shine light on understanding the molecular mechanisms of FGF signaling, but also will provide guidance on developing novel cancer therapeutic strategies based on inhibiting aberrant FRS2-mediated signals.

10. Closing Remarks

Although the four FGFR kinases are highly homologous, the signals elicited by them are both overlapping and in context highly specific. However, the determinants for the specificities of FGF signaling are largely unknown. As the major adaptor for the FGFR kinases, FRS2 has multiple tyrosine phosphorylation sites that, upon phosphorylation by receptor tyrosine kinases, constitute binding sites for downstream signaling molecules. Although GRB2 and SHP2 are the two signaling molecules that clearly bind to FRS2, whether phosphotyrosines of FRS2 also bind to and organize other downstream molecules remains an open question. Whether all GRB2 or SHP2 binding sites bind to GRB2 or SHP2 proteins equally and relay the same signals to downstream effectors is also unknown. Finally, whether the two FRS2α isotypes mediate the same signaling pathways for the FGFR kinases or for other receptor tyrosine kinases is also poorly understood. The detailed molecular mechanisms of how aberrant signals mediated by FRS2α contribute to diseases, including cancers, also need to be completely characterized. Answers to these key issues related to FRS2-mediated signals will not only be important for understanding the molecular mechanism by which FGF regulates a wide spectrum of biological processes in a spatiotemporal specific manner, but also will provide guidance to development of novel therapeutic strategies based on targeting FRS2-mediated signaling pathways.

References

1. McKeehan, W. L., Wang, F., and Luo, Y. *The fibroblast growth factor (FGF) signaling complex. Handbook of Cell Signaling,* 2dn ed., Academic/Elsevier Press, New York (2009).
2. McKeehan, W. L., Wang, F., and Kan, M. The heparan sulfate-fibroblast growth factor family: Diversity of structure and function. *Prog Nucleic Acid Res Mol Biol* **59**, 135–176 (1998).
3. Wang, F., McKeehan, K., Yu, C., and McKeehan, W. L. Fibroblast growth factor receptor 1 phosphotyrosine 766: Molecular target for prevention of progression of prostate tumors to malignancy. *Cancer Res* **62**, 1898–1903 (2002).

4. Powers, C. J., McLeskey, S. W., and Wellstein, A. Fibroblast growth factors, their receptors and signaling. *Endocr Relat Cancer* **7**, 165–197 (2000).
5. Rabin, S. J., Cleghon, V., and Kaplan, D. R. SNT, a differentiation-specific target of neurotrophic factor-induced tyrosine kinase activity in neurons and PC12 cells. *Mol Cell Biol* **13**, 2203–2213 (1993).
6. Wang, J. K., Xu, H., Li, H. C., and Goldfarb, M. Broadly expressed SNT-like proteins link FGF receptor stimulation to activators of Ras. *Oncogene* **13**, 721–729 (1996).
7. Kouhara, H., Hadari, Y. R., Spivak-Kroizman, T., Schilling, J., *et al.* A lipid-anchored Grb2-binding protein that links FGF-receptor activation to the Ras/MAPK signaling pathway. *Cell* **89**, 693–702 (1997).
8. Xu, H., Lee, K. W., and Goldfarb, M. Novel recognition motif on fibroblast growth factor receptor mediates direct association and activation of SNT adapter proteins. *J Biol Chem* **273**, 17987–17990 (1998).
9. McDougall, K., Kubu, C., Verdi, J. M., and Meakin, S. O. Developmental expression patterns of the signaling adapters FRS-2 and FRS-3 during early embryogenesis. *Mech Dev* **103**, 145–148 (2001).
10. Dhalluin, C., Yan, K., Plotnikova, O., Lee, K. W., Zeng, L., *et al.* Structural basis of SNT PTB domain interactions with distinct neurotrophic receptors. *Mol Cell* **6**, 921–929 (2000).
11. Dhalluin, C., Yan, K. S., Plotnikova, O., Zeng, L., *et al.* 1H, 13C and 15N resonance assignments of the SNT PTB domain in complex with FGFR1 peptide. *J Biomol NMR* **18**, 371–372 (2000).
12. Yan, K. S., Kuti, M., Yan, S., Mujtaba, S., *et al.* FRS2 PTB domain conformation regulates interactions with divergent neurotrophic receptors. *J Biol Chem* **277**, 17088–17094 (2002).
13. Xu, H., and Goldfarb, M. Multiple effector domains within SNT1 coordinate ERK activation and neuronal differentiation of PC12 cells. *J Biol Chem* **276**, 13049–13056 (2001).
14. Zhou, L., McDougall, K., Kubu, C. J., Verdi, J. M., *et al.* Genomic organization and comparative sequence analysis of the mouse and human FRS2, FRS3 genes. *Mol Biol Rep* **30**, 15–25 (2003).
15. Melillo, R. M., Santoro, M., Ong, S. H., Billaud, M., *et al.* Docking protein frs2 links the protein tyrosine kinase ret and its oncogenic forms with the mitogen-activated protein kinase signaling cascade. *Mol Cell Biol* **21**, 4177–4187 (2001).
16. Cavallaro, U., Niedermeyer, J., Fuxa, M., and Christofori, G. N-CAM modulates tumour-cell adhesion to matrix by inducing FGF-receptor signalling. *Nat Cell Biol* **3**, 650–657 (2001).

17. Ong, S. H., Hadari, Y. R., Gotoh, N., Guy, G. R., *et al.* Stimulation of phosphatidylinositol 3-kinase by fibroblast growth factor receptors is mediated by coordinated recruitment of multiple docking proteins. *Proc Natl Acad Sci USA* **98**, 6074–6079 (2001).
18. Ong, S. H., Goh, K. C., Lim, Y. P., Low, B. C., *et al.* Suc1-associated neurotrophic factor target (SNT) protein is a major FGF- stimulated tyrosine phosphorylated 90-kDa protein which binds to the SH2 domain of GRB2. *Biochem Biophys Res Commun* **225**, 1021–1026 (1996).
19. Ong, S. H., Lim, Y. P., Low, B. C., and Guy, G. R. SHP2 associates directly with tyrosine phosphorylated p90 (SNT) protein in FGF-stimulated cells. *Biochem Biophys Res Commun* **238**, 261–266 (1997).
20. Hadari, Y. R., Gotoh, N., Kouhara, H., Lax, I., *et al.* Critical role for the docking-protein FRS2 alpha in FGF receptor-mediated signal transduction pathways. *Proc Natl Acad Sci USA* **98**, 8578–8583 (2001).
21. Wong, A., Lamothe, B., Lee, A., Schlessinger, J., *et al.* FRS2 alpha attenuates FGF receptor signaling by Grb2-mediated recruitment of the ubiquitin ligase Cbl. *Proc Natl Acad Sci USA* **99**, 6684–6689 (2002).
22. Hadari, Y. R., Kouhara, H., Lax, I., and Schlessinger, J. Binding of Shp2 tyrosine phosphatase to FRS2 is essential for fibroblast growth factor-induced PC12 cell differentiation. *Mol Cell Bio* **18**, 3966–3973 (1998).
23. Kontaridis, M. I., Liu, X., Zhang, L., and Bennett, A. M. Role of SHP-2 in fibroblast growth factor receptor-mediated suppression of myogenesis in C2C12 myoblasts. *Mol Cell Biol* **22**, 3875–3891 (2002).
24. Cross, M. J., Lu, L., Magnusson, P., Nyqvist, D., *et al.* The Shb adaptor protein binds to tyrosine 766 in the FGFR-1 and regulates the Ras/MEK/MAPK pathway via FRS2 phosphorylation in endothelial cells. *Mol Bio Cell* **13**, 2881–2893 (2002).
25. Krejci, P., Masri, B., Salazar, L., Farrington-Rock, C., *et al.* Bisindolylmaleimide I suppresses fibroblast growth factor-mediated activation of Erk MAP kinase in chondrocytes by preventing Shp2 association with the Frs2 and Gab1 adaptor proteins. *J Biol Chem* **282**, 2929–2936 (2007).
26. Ridyard, M. S., and Robbins, S. M. Fibroblast growth factor-2-induced signaling through lipid raft-associated fibroblast growth factor receptor substrate 2 (FRS2). *J Biol Chem* **278**, 13803–13809 (2003).
27. Tomasovic, A., Traub, S., and Tikkanen, R. Molecular networks in FGF signaling: flotillin-1 and cbl-associated protein compete for the binding to fibroblast growth factor receptor substrate 2. *PloS one* **7**, e29739 (2012).

28. Zhang, Y., Lin, Y., Bowles, C., and Wang, F. Direct cell cycle regulation by the Fibroblast Growth Factor Receptor (FGFR) Kinase through Phosphorylation-dependent Release of Cks1 from FGFR Substrate 2. *J Biol Chem* **279**, 55348–55354 (2004).
29. Harada, A., Katoh, H., and Negishi, M. Direct interaction of Rnd1 with FRS2 beta regulates Rnd1-induced down-regulation of RhoA activity and is involved in fibroblast growth factor-induced neurite outgrowth in PC12 cells. *J Biol Chem* **280**, 18418–18424 (2005).
30. Lin, X., Zhang, Y., Liu, L., McKeehan, W. L., *et al.* FRS2alpha is essential for the fibroblast growth factor to regulate the mTOR pathway and autophagy in mouse embryonic fibroblasts. *Int J Biol Sci* **7**, 1114–1121 (2011).
31. Chen, P. Y., and Friesel, R. FGFR1 forms an FRS2-dependent complex with mTOR to regulate smooth muscle marker gene expression. *Biochem Biophys Res Commun* **382**, 424–429 (2009).
32. Kao, S., Jaiswal, R. K., Kolch, W., and Landreth, G. E. Identification of the mechanisms regulating the differential activation of the mapk cascade by epidermal growth factor and nerve growth factor in PC12 cells. *J Biol Chem* **276**, 18169–18177 (2001).
33. Wang, Z., Dela Cruz, R., Ji, F., Guo, S., *et al.* G(i)alpha proteins exhibit functional differences in the activation of ERK1/2, Akt and mTORC1 by growth factors in normal and breast cancer cells. *Cell Commun Signal: CCS* **12**, 10 (2014).
34. Gotoh, N., Laks, S., Nakashima, M., Lax, I., *et al.* FRS2 family docking proteins with overlapping roles in activation of MAP kinase have distinct spatial-temporal patterns of expression of their transcripts. *FEBS Lett* **564**, 14–18 (2004).
35. Hryciw, T., MacDonald, J. I., Phillips, R., Seah, C., *et al.* The fibroblast growth factor receptor substrate 3 adapter is a developmentally regulated microtubule-associated protein expressed in migrating and differentiated neurons. *J Neurochem* **112**, 924–939 (2010).
36. Zhou, L., Talebian, A., and Meakin, S. O. The signaling adapter, FRS2, facilitates neuronal branching in primary cortical neurons via both Grb2- and Shp2-dependent mechanisms. *J Mol Neurosci* **55**, 663–677 (2014).
37. Dhalluin, C., Yan, S. K., Plotnikova, O., Lee, W. K., *et al.* Structural basis of SNT PTB domain interactions with distinct neurotrophic receptors. *Mol Cell* **6**, 921–929 (2000).

38. Dhe-Paganon, S., Ottinger, E. A., Nolte, R. T., Eck, M. J., *et al.* Crystal structure of the pleckstrin homology-phosphotyrosine binding (PH-PTB) targeting region of insulin receptor substrate 1. *Proc Nat Acad Sci USA* **96**, 8378–8383 (1999).
39. Eck, M. J., Dhe-Paganon, S., Trub, T., Nolte, R. T., *et al.* Structure of the IRS-1 PTB domain bound to the juxtamembrane region of the insulin receptor. *Cell* **85**, 695–705 (1996).
40. Nolte, R. T., Eck, M. J., Schlessinger, J., Shoelson, S. E., *et al.* Crystal structure of the PI 3-kinase p85 amino-terminal SH2 domain and its phosphopeptide complexes. *Nat Struct Biol* **3**, 364–374 (1996).
41. Burgar, H. R., Burns, H. D., Elsden, J. L., Lalioti, M. D., *et al.* Association of the signaling adaptor FRS2 with fibroblast growth factor receptor 1 (Fgfr1) is mediated by alternative splicing of the juxtamembrane domain. *J Biol Chem* **277**, 4018–4023 (2002).
42. Paterno, G. D., Ryan, P. J., Kao, K. R., *et al.* The VT+ and VT- isoforms of the fibroblast growth factor receptor type 1 are differentially expressed in the presumptive mesoderm of Xenopus embryos and differ in their ability to mediate mesoderm formation. *J Biol Chem* **275**, 9581–9586 (2000).
43. Hoch, R. V., and Soriano, P. Context-specific requirements for Fgfr1 signaling through Frs2 and Frs3 during mouse development. *Develop* **133**, 663–673 (2006).
44. Zhang, Y., McKeehan, K., Lin, Y., Zhang, J., *et al.* Fibroblast growth factor receptor 1 (FGFR1) tyrosine phosphorylation regulates binding of FGFR substrate 2alpha (FRS2alpha) but not FRS2 to the receptor. *Mol Endocrinol* **22**, 167–175 (2008).
45. Wang, F. Cell- and receptor isotype-specific phosphorylation of SNT1 by fibroblast growth factor receptor tyrosine kinases. *In vitro Cell Dev Biol Anim* **38**, 178–183 (2002).
46. Wu, Y., Chen, Z., and Ullrich, A. EGFR and FGFR signaling through FRS2 is subject to negative feedback control by ERK1/2. *Biol Chem* **384**, 1215–1226 (2003).
47. Zeng, G., and Meakin, S. O. Overexpression of the signaling adapter FRS2 reconstitutes the cell cycle deficit of a nerve growth factor non-responsive TrkA receptor mutant. *J Neurochem* **81**, 820–831 (2002).
48. Meakin, S. O., MacDonald, J. I., Gryz, E. A., Kubu, C. J., *et al.* The signaling adapter FRS-2 competes with Shc for binding to the nerve growth factor receptor TrkA. A model for discriminating proliferation and differentiation. *J Biol Chem* **274**, 9861–9870 (1999).

49. Zeng, L., Kuti, M., Mujtaba, S., and Zhou, M. M. Structural insights into FRS2alpha PTB domain recognition by neurotrophin receptor TrkB. *Proteins* **82**, 1534–1541 (2014).
50. Ong, S. H., Guy, G. R., Hadari, Y. R., Laks, S., *et al.* FRS2 proteins recruit intracellular signaling pathways by binding to diverse targets on fibroblast growth factor and nerve growth factor receptors. *Mol Cell Biol* **20**, 979–989 (2000).
51. Easton, J. B., Moody, N. M., Zhu, X., and Middlemas, D. S. Brain-derived neurotrophic factor induces phosphorylation of fibroblast growth factor receptor substrate 2. *J Biol Chem* **274**, 11321–11327 (1999).
52. Easton, J. B., Royer, A. R., and Middlemas, D. S. The protein tyrosine phosphatase, Shp2, is required for the complete activation of the RAS/MAPK pathway by brain-derived neurotrophic factor. *J Neurochem* **97**, 834–845 (2006).
53. Dixon, S. J., MacDonald, J. I., Robinson, K. N., Kubu, C. J., *et al.*, Trk receptor binding and neurotrophin/fibroblast growth factor (FGF)-dependent activation of the FGF receptor substrate (FRS)-3. *Biochim Biophys Acta* **1763**, 366–380 (2006).
54. Yamada, M., Suzuki, K., Mizutani, M., Asada, A., *et al.* Analysis of tyrosine phosphorylation-dependent protein-protein interactions in TrkB-mediated intracellular signaling using modified yeast two-hybrid system. *J Biochem (Tokyo)* **130**, 157–165 (2001).
55. Kurokawa, K., Iwashita, T., Murakami, H., Hayashi, H., *et al.* Identification of SNT/FRS2 docking site on RET receptor tyrosine kinase and its role for signal transduction. *Oncogene* **20**, 1929–1938 (2001).
56. Lundgren, T. K., Luebke, M., Stenqvist, A., and Ernfors, P. Differential membrane compartmentalization of Ret by PTB-adaptor engagement. *FEBS J* **275**, 2055–2066 (2008).
57. Lundgren, T. K., Stenqvist, A., Scott, R. P., Pawson, T., *et al.* Cell migration by a FRS2-adaptor dependent membrane relocation of ret receptors. *J Cell Biochem* **104**, 879–894 (2008).
58. Stoletov, K. V., Ratcliffe, K. E., and Terman, B. I. Fibroblast growth factor receptor substrate 2 participates in vascular endothelial growth factor-induced signaling. *Faseb J* **16**, 1283–1285 (2002).
59. Chen, P. Y., Qin, L., Zhuang, Z. W., Tellides, G., *et al.* The docking protein FRS2alpha is a critical regulator of VEGF receptors signaling. *Proc Natl Acad Sci USA* **111**, 5514–5519 (2014).

60. Delahaye, L., Rocchi, S., and Van Obberghen, E. Potential involvement of FRS2 in insulin signaling. *Endocrinology* **141**, 621–628 (2000).
61. Chen, P. Y., Simons, M., and Friesel, R. FRS2 via fibroblast growth factor receptor 1 is required for platelet-derived growth factor receptor beta-mediated regulation of vascular smooth muscle marker gene expression. *J Biol Chem* **284**, 15980–15992 (2009).
62. Lax, I., Wong, A., Lamothe, B., Lee, A., *et al.* The docking protein FRS2alpha controls a MAP kinase-mediated negative feedback mechanism for signaling by FGF receptors. *Mol Cell* **10**, 709–719 (2002).
63. Zhou, W., Feng, X., Wu, Y., Benge, J., *et al.* FGF-receptor substrate 2 functions as a molecular sensor integrating external regulatory signals into the FGF pathway. *Cell Res* **19**, 1165–1177 (2009).
64. Avery, A. W., Figueroa, C., and Vojtek, A. B. UNC-51-like kinase regulation of fibroblast growth factor receptor substrate 2/3. *Cell Signal* **19**, 177–184 (2006).
65. Li, X., Brunton, V. G., Burgar, H. R., Wheldon, L. M., and Heath, J. K. FRS2-dependent SRC activation is required for fibroblast growth factor receptor-induced phosphorylation of Sprouty and suppression of ERK activity. *J Cell Sci* **117**, 6007–6017 (2004).
66. Huang, L., Gotoh, N., Zhang, S., Shibuya, M., *et al.* SNT-2 interacts with ERK2 and negatively regulates ERK2 signaling in response to EGF stimulation. *Biochem Biophys Res Commun* **324**, 1011–1017 (2004).
67. Minegishi, Y., Shibagaki, Y., Mizutani, A., Fujita, K., *et al.* Adaptor protein complex of FRS2beta and CIN85/CD2AP provides a novel mechanism for ErbB2/HER2 protein downregulation. *Cancer Sci* **104**, 345–352 (2013).
68. Huang, L., Watanabe, M., Chikamori, M., Kido, Y., *et al.* Unique role of SNT-2/FRS2beta/FRS3 docking/adaptor protein for negative regulation in EGF receptor tyrosine kinase signaling pathways. *Oncogene* **25**, 6457–6466 (2006).
69. Haling, J. R., Wang, F., and Ginsberg, M. H. Phosphoprotein enriched in astrocytes 15 kDa (PEA-15) reprograms growth factor signaling by inhibiting threonine phosphorylation of fibroblast receptor substrate 2alpha. *Mol Biol Cell* **21**, 664–673 (2010).
70. Gotoh, N., Ito, M., Yamamoto, S., Yoshino, I., *et al.* Tyrosine phosphorylation sites on FRS2alpha responsible for Shp2 recruitment are critical for induction of lens and retina. *Proc Natl Acad Sci USA* **101**, 17144–17149 (2004).

71. Yamamoto, S., Yoshino, I., Shimazaki, T., Murohashi, M., *et al.* Essential role of Shp2-binding sites on FRS2{alpha} for corticogenesis and for FGF2-dependent proliferation of neural progenitor cells. *Proc Natl Acad Sci USA* (2005).
72. Kameda, Y., Ito, M., Nishimaki, T., and Gotoh, N. FRS2alpha is required for the separation, migration, and survival of pharyngeal-endoderm derived organs including thyroid, ultimobranchial body, parathyroid, and thymus. *Developmental Dynamics: An Official Publication of the American Association of Anatomists* **238**, 503–513 (2009).
73. Kameda, Y., Ito, M., Nishimaki, T., and Gotoh, N. FRS2alpha 2F/2F mice lack carotid body and exhibit abnormalities of the superior cervical sympathetic ganglion and carotid sinus nerve. *Dev Biol* **314**, 236–247 (2008).
74. Lin, Y., Zhang, J., Zhang, Y., and Wang, F. Generation of an Frs2alpha conditional null allele. *Genesis* **45**, 554–559 (2007).
75. Li, H., Tao, C., Cai, Z., Hertzler-Schaefer, K., *et al.* Frs2alpha and Shp2 signal independently of Gab to mediate FGF signaling in lens development. *J Cell Sci* **127**, 571–582 (2014).
76. Ono, K., Kita, T., Sato, S., O'Neill, P., *et al.* FGFR1-Frs2/3 signalling maintains sensory progenitors during inner ear hair cell formation. *PLoS Genetics* **10**, e1004118 (2014).
77. Zhang, J., Lin, Y., Zhang, Y., Lan, Y., *et al.* Frs2alpha-deficiency in cardiac progenitors disrupts a subset of FGF signals required for outflow tract morphogenesis. *Development* **135**, 3611–3622 (2008).
78. Zhang, J., Chang, J. Y., Huang, Y., Lin, X., *et al.* The FGF-BMP signaling axis regulates outflow tract valve primordium formation by promoting cushion neural crest cell differentiation. *Circ Res* **107**, 1209–1219 (2010).
79. Zhang, J., Liu, J., Huang, Y., Chang, J. Y., *et al.* FRS2alpha-mediated FGF signals suppress premature differentiation of cardiac stem cells through regulating autophagy activity. *Circ Res* **110**, e29–39 (2012).
80. Chen, P. Y., Qin, L., Barnes, C., Charisse, K., *et al.* FGF regulates TGF-beta signaling and endothelial-to-mesenchymal transition via control of let-7 miRNA expression. *Cell Rep* **2**, 1684–1696 (2012).
81. Lin, Y., Liu, G., Zhang, Y., Hu, Y. P., *et al.* Fibroblast growth factor receptor 2 tyrosine kinase is required for prostatic morphogenesis and the acquisition of strict androgen dependency for adult tissue homeostasis. *Development* **134**, 723–734 (2007).

82. Yan, G., Fukabori, Y., McBride, G., Nikolaropolous, S., *et al.* Exon switching and activation of stromal and embryonic fibroblast growth factor (FGF)-FGF receptor genes in prostate epithelial cells accompany stromal independence and malignancy. *Mol Cell Biol* **13**, 4513–4522 (1993).
83. Feng, S., Wang, F., Matsubara, A., Kan, M., *et al.* Fibroblast growth factor receptor 2 limits and receptor 1 accelerates tumorigenicity of prostate epithelial cells. *Cancer Res* **57**, 5369–5378 (1997).
84. Zhang, Y., Zhang, J., Lin, Y., Lan, Y., *et al.* Role of epithelial cell fibroblast growth factor receptor substrate 2{alpha} in prostate development, regeneration and tumorigenesis. *Development* **135**, 775–784 (2008).
85. Xu, B., Yang, L., and Hinton, B. T. The Role of fibroblast growth factor receptor substrate 2 (FRS2) in the regulation of two activity levels of the components of the extracellular signal-regulated kinase (ERK) pathway in the mouse epididymis. *Biol Reprod* **89**, 48 (2013).
86. Hama, J., Xu, H., Goldfarb, M., and Weinstein, D. C. SNT-1/FRS2alpha physically interacts with Laloo and mediates mesoderm induction by fibroblast growth factor. *Mech Dev* **109**, 195–204 (2001).
87. Akagi, K., Kyun Park, E., Mood, K., and Daar, I. O. Docking protein SNT1 is a critical mediator of fibroblast growth factor signaling during Xenopus embryonic development. *Developmental Dynamics: An Official Publication of the American Association of Anatomists* **223**, 216–228 (2002).
88. Kusakabe, M., Masuyama, N., Hanafusa, H., and Nishida, E. Xenopus FRS2 is involved in early embryogenesis in cooperation with the Src family kinase Laloo. *EMBO Rep* **2**, 727–735 (2001).
89. Mood, K., Friesel, R., and Daar, I. O. SNT1/FRS2 mediates germinal vesicle breakdown induced by an activated FGF receptor1 in Xenopus oocytes. *J Biol Chem* **277**, 33196–33204 (2002).
90. Lo, T. W., Bennett, D. C., Goodman, S. J., and Stern, M. J. Caenorhabditis elegans fibroblast growth factor receptor signaling can occur independently of the multi-substrate adaptor FRS2. *Genetics* **185**, 537–547 (2010).
91. Zhang, K., Chu, K., Wu, X., Gao, H., *et al.* Amplification of FRS2 and activation of FGFR/FRS2 signaling pathway in high-grade liposarcoma. *Cancer Res* **73**, 1298–1307 (2013).
92. Kakazu, N., Yamane, H., Miyachi, M., Shiwaku, K., *et al.* Identification of the 12q15 amplicon within the homogeneously staining regions in the embryonal rhabdomyosarcoma cell line RMS-YM. *Cytogenet Genome Res* **142**, 167–173 (2014).

93. De Martino, M. C., Al Ghuzlan, A., Aubert, S., Assie, G., *et al.* Molecular screening for a personalized treatment approach in advanced adrenocortical cancer. *J Clin Endocrinol Metab* **98**, 4080–4088 (2013).
94. Fischer, U., Keller, A., Leidinger, P., Deutscher, S., *et al.* A different view on DNA amplifications indicates frequent, highly complex, and stable amplicons on 12q13-21 in glioma. *Mol Cancer Res: MCR* **6**, 576–584 (2008).
95. Wang, X., Asmann, Y. W., Erickson-Johnson, M. R., Oliveira, J. L., *et al.* High-resolution genomic mapping reveals consistent amplification of the fibroblast growth factor receptor substrate 2 gene in well-differentiated and dedifferentiated liposarcoma. *Genes, Chromosomes & Cancer* **50**, 849–858 (2011).
96. Chen, Y., McGee, J., Chen, X., Doman, T. N., *et al.* Identification of druggable cancer driver genes amplified across TCGA datasets. *PloS One* **9**, e98293 (2014).
97. Liu, J, P. You, G. Chen, X. Fu, X. Zeng, *et al.*, Hyperactivated FRS2α-mediated signaling in prostate cancer cells promotes tumor angiogenesis and predicts poor clinical outcome of patients. *Oncogene* **35**, 1750–1759 (2016).
98. Valencia, T., Joseph, A., Kachroo, N., Darby, S., *et al.* Role and expression of FRS2 and FRS3 in prostate cancer. *BMC cancer* **11**, 484 (2011).
99. Ren, J., Huang, H. J., Gong, Y., Yue, S., *et al.* MicroRNA-206 suppresses gastric cancer cell growth and metastasis. *Cell Biosci* **4**, 26 (2014).
100. Zhang, H., Hylander, B. L., LeVea, C., Repasky, E. A., *et al.* Enhanced FGFR signalling predisposes pancreatic cancer to the effect of a potent FGFR inhibitor in preclinical models. *Br J Cancer* **110**, 320–329 (2014).
101. Manuvakhova, M., Thottassery, J. V., Hays, S., Qu, Z., *et al.* Expression of the SNT-1/FRS2 phosphotyrosine binding domain inhibits activation of MAP kinase and PI3-kinase pathways and antiestrogen resistant growth induced by FGF-1 in human breast carcinoma cells. *Oncogene* (2006).
102. Roccato, E., Miranda, C., Ranzi, V., Gishizki, M., *et al.* Biological activity of the thyroid TRK-T3 oncogene requires signalling through Shc. *Br J Cancer* **87**, 645–653 (2002).
103. Ranzi, V., Meakin, S. O., Miranda, C., Mondellini, P., *et al.* The signaling adapters fibroblast growth factor receptor substrate 2 and 3 are activated by the thyroid TRK oncoproteins. *Endocrinology* **144**, 922–928 (2003).

104. Chikamori, M., Fujimoto, J., Tokai-Nishizumi, N., and Yamamoto, T. Identification of multiple SNT-binding sites on NPM-ALK oncoprotein and their involvement in cell transformation. *Oncogene* **26**, 2950–2954 (2007).
105. Sims-Lucas, S., Cusack, B., Eswarakumar, V. P., Zhang, J., *et al.* Independent roles of Fgfr2 and Frs2alpha in ureteric epithelium. *Development* **138**, 1275–1280 (2011).
106. Iejima, D., Minegishi, Y., Takenaka, K., Siswanto, A., *et al.* FRS2beta, a potential prognostic gene for non-small cell lung cancer, encodes a feedback inhibitor of EGF receptor family members by ERK binding. *Oncogene* **29**, 3087–3099 (2010).

Chapter 4

Fibroblast Growth Factor Signaling, Endothelial Homeostasis, and Endothelial Cell to Mesenchymal Transition

Pei-Yu Chen* and Michael Simons*,†,‡

Abstract

Fibroblast growth factors and their receptors play an important, if underappreciated roles in endothelial cell biology. Recent studies suggested their involvement in regulation of vascular permeability, stability and homeostasis. Most intriguingly, endothelial FGF signaling emerged as the key regulator of transforming growth factor beta (TGFβ) activity and endothelial-to-mesenchymal transition. These and other facts of endothelial FGF biology are explored in this Chapter.

*Yale Cardiovascular Research Center, Section of Cardiovascular Medicine, Department of Internal Medicine, Yale University School of Medicine, New Haven, CT, 06520, USA.

†Department of Cell Biology, Yale University School of Medicine, New Haven, CT, 06520, USA.

‡Corresponding author: E-mail: michael.simons@yale.edu

1. Overview of Fibroblast Growth Factor Signaling

Fibroblast growth factors (FGFs) and their receptors are part of a large family of signaling molecules that are highly conserved and found in organisms ranging from *Caenorhabditis elegans* to humans. FGFs exert their activity via family of four related receptor tyrosine kinases (RTKs) designated FGFR1-FGFR4. In addition, FGF signaling is regulated by a number of accessory proteins including α- and β-Klotho and syndecans. The FGF/FGFR system has been implicated in a variety of physiological and pathological conditions, including embryonic development, tissue growth and remodeling, inflammation, tumor growth, vascularization, and wound healing.[1] Here we will consider its role in maintenance of endothelial homeostasis and regulation of endothelial-to-mesenchymal (EndMT) transition.

As with most RTKs, FGFR dimerization by an FGF ligand leads to activation of the intracellular kinase and autophosphorylation tyrosine residues in their cytoplasmic domain, thereby creating docking sites for a variety of intracellular adaptor proteins.[2] In the case of FGFR1, the most prevalent FGF receptor, these include Tyrosine 463 that is responsible for Crk (CT10 (chicken tumor virus number 10) regulator of kinase) binding,[3] Tyrosine 653/654 that are critical for FGFR1 tyrosine kinase activity, and Tyrosine 766 that is responsible for PLCγ (phospholipase Cγ) and Shb (SH2 Domain Containing Adaptor Protein B) binding.[4–6]

FGFRs activate several downstream signaling cascades including the MAPK (Ras/Raf-1/MEK/ERK1/2), PLCγ, and PI3K (phosphatidylinositide 3-kinase) pathway (Fig. 1). These will be discussed in turn. FGFRs activate the MAPK pathway via a unique membrane-anchored docking protein fibroblast growth factor receptor substrate 2 (FRS2). There are two family members in mammals (FRS2α and FRS2β), one homolog in Drosophila (dFRS2), one homolog in Xenopus (xFRS2), and two homologs in zebrafish. In mammals, FRS2α and FRS2β have a similar structural architecture but only share 48% amino acid sequence identity and occupy separate chromosomal loci.[7] FRS2α is broadly expressed in adult and fetal tissues, while FRS2β expression is more restricted to cells of neuronal origin.[8]

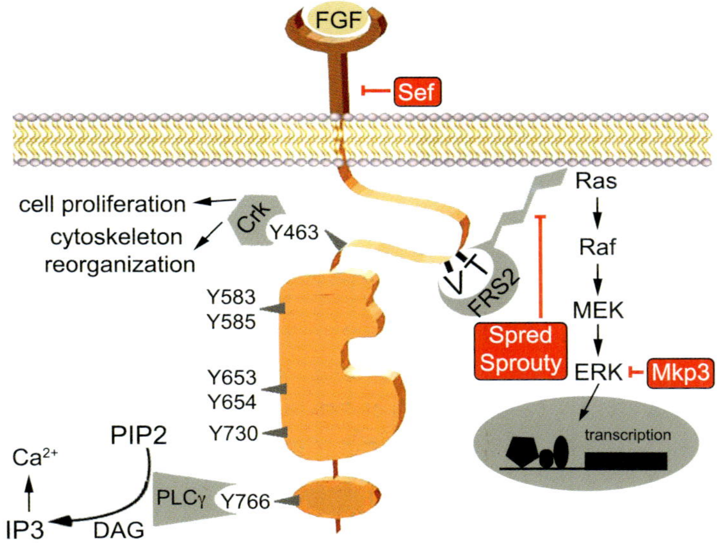

Fig. 1. FGF mediated signal transduction pathways and it's feedback regulators. The three main signaling pathways downstream of FGF receptor activation are the Ras-MAP kinase, PLCγ, and PI3K pathways. Sprouty, Sef (similar expression to FGF), and MKP3 (MAP Kinase Phosphatase 3) provide a negative-feedback mechanism to reduce or terminate FGF signaling.

Unlike other adaptor proteins that bind to RTKs in a ligand-dependent manner, FRS2α associates with the FGFRs in a ligand-independent fashion via an unusual N-terminal amino acid sequence.[9] Tyrosine phosphorylation of FRS2α by an activated FGFR on 6 distinct residues creates binding sites for Grb2 (Growth Factor Receptor Bound Protein 2) and Shp2 (Src-homology 2 domain-containing phosphatase 2) adaptor proteins that are critical for activation of the MAPK (Ras/Raf-1/MEK/ERK1/2) signaling cascade.[10] Activated ERK1/2, a serine/threonine kinase, is then able to phosphorylate and modify the activity of a number of transcription factors. Among these are the ETS (E-twenty six) family transcription factors that use a winged helix-turn-helix protein fold as a DNA binding domain. They have been shown to be key effectors of FGF signaling regulating expression of a number critical endothelial genes including VEGFR2 (Vascular Endothelial Growth Factor Receptor 2).[11–13]

Activation of the PLCγ pathway occurs when its Src homology 2 (SH2) domain binds the phosphorylated tyrosine residue (Tyr766) of FGFR1. Thus activated PLCγ hydrolyses phosphatidylinositol-4,5-diphospate (PIP2) to inositol-1,4,5-triphosphate (IP3) and diacylglycerol (DAG). IP3 is able to induce Ca^{2+} release from storage, while DAG activates protein kinase C (PKC), which in turn is able to phosphorylate Raf and activate the MAPK pathway.[14]

The PI3K (phosphoinositol-3-kinase) pathway is activated by Gab1 (GRB2 Associated Binding Protein 1) binding to phosphorylated FRS2α through Grb2.[15] This induces PI3K to phosphorylate PIP2 to generate phosphatidylinositol-3,4,5-triphospate (PIP3) which induces serine/threonine kinase Akt activation. Several other feedback regulators, such as Sprouty, Sef (similar expression to FGF), and MKP3 (MAP Kinase Phosphatase 3), are activated through the FGF signaling cascade.

2. The Biology of Endothelial FGF Signaling

FGFRs are widely expressed. In endothelial cells, FGFR1 is the most abundant FGFR-encoding mRNA, FGFR2 mRNA expression is almost undetectable and FGFR3 and FGFR4 are expressed at fairly low levels.[16] While all FGFRs have been knocked out in mouse studies, the phenotypes of these global knockouts have not been particularly informative with regard to vascular development either due to very early embryonic lethality occurring prior to critical stages of blood vessel development (FGFR1 and FGFR2),[17,18] or to lack of an obvious vascular phenotype in viable mice (FGFR3 and FGFR4).[19–21]

Targeted disruption of the *Frs2α* gene also causes severe impairment in mouse development, resulting in lethality at embryonic day E7.0–E7.5.[22] *Frs2β* knockout mice have not yet been described. Clearly, while a great deal is known about the expression, activity and function of FGF signaling in endothelial cells *in vitro*, there is much less knowledge about the expression and function of FGF signaling in endothelial cells *in vivo*.

At the same time, progress has been made in elucidating the role of FGF signaling *in vivo*. The loss-of-function study in adult mice using a dominant negative FGFR1 (dnFGFR1) construct demonstrated a dramatic reduction in endothelial cell-cell junctions and a gradual loss of vascular integrity leading to increased permeability and hemorrhages.[13,23] Another important role of endothelial FGF signaling is the maintenance of VEGFR2 expression. FGF2 has been shown to markedly increased VEGFR2 expression at both mRNA and protein levels in endothelial cells.[24,25] Further studies showed that FGF regulates the VEGFR2 enhancer through an FOX-ETS motif in an ERK1/2-dependent manner.[13] An endothelium-specific deletion of both FGFR1/FGFR2 produced viable mice with no vascular developmental defects.[26] However, there was a significant impairment of angiogenic response after skin or eye injury with the former leading to delayed cutaneous wound healing.[26]

More recently, the loss of endothelial FGF signaling has been linked with neointima formation. An endothelial cell-specific knockout of either *Fgfr1* or *Frs2α* enhanced neointima formation both in transplant arteriosclerosis and lipid-induced arteriosclerosis model.[16,27,28] This will be discussed in greater detail below. Finally, FGF signaling has been shown to be required for the maintenance of both blood and lymphatic endothelial cells identity through FGF-Ras-MAPK signaling.[29,30]

As these results indicate, FGFs play an important and poorly understood role in the vasculature and much work is still required to fully elucidate their contributions. Its various activities will be examined in turn.

3. FGF-dependent Regulation of Vascular Integrity

Endothelial cells play an essential role in the maintenance of vascular integrity and FGF signaling is a prominent contributor. This is illustrated by inhibition of FGF signaling achieved either by a systemic expression of a soluble FGF receptor trap (sFGFR1-IIIc) that binds all FGFs available to activate various FGFRs, or an endothelial-specific expression of a truncated form of FGFR1 missing its cytoplasmic domain. This causes a loss of endothelial cell-cell adhesion leading to disruption of the endothelial barrier and impairment of vascular integrity.[23]

At the molecular level, the lack of endothelial FGF signaling leads to dissociation of p120-catenin from VE-cadherin due to increased phosphorylation of the latter on Y658, the site involved in VE-cadherin-p120-catenin binding.[31] This results in VE-cadherin-p120-catenin dissociation and the loss of VE-cadherin from adherence junctions setting leading to the loss of endothelial barrier function that can progress to a frank loss of vascular integrity and hemorrhage.

This increased VE-cadherin Y658 phosphorylation has been traced to decrease in the phosphatase SHP2 expression that is FGF-dependent. Indeed, increased permeability induced by decreased FGF signaling can be fully reversed by SHP2 overexpression.[32]

4. FGF-TGFβ Cross-talk and Endothelial-to-Mesenchymal Transition

One of the consequences of impaired endothelial FGF signaling input is the development of endothelial-to-mesenchymal transition (EndMT). EndMT was originally described as an embryonic phenomenon that plays a prominent role in cardiac valve formation.[33,34] *In vitro* EndMT is characterized by the loss of cell-cell adhesions and changes in cell polarity, leading to cuboidal endothelial cells acquiring a spindle-shaped morphology. The expression of endothelial cell markers, such as VE-cadherin and CD31 is decreased but not fully extinguished, whilst the expression of mesenchymal cell markers, such as α-smooth muscle actin (α-SMA) and SM-calponin is increased. Functionally endothelial cells undergoing EndMT acquire myofibroblast-like characteristics including the ability contract, enhanced cell motility, and increased extracellular matrix production.

Cardiac development aside, more recent studies have shown that EndMT occurs in a number of disease entities including tissue fibrosis,[35] neointima formation,[16,27,36] atherosclerosis,[28] cerebral cavernous malformation (CCM),[37] and pulmonary hypertension[38,39] among others. Given the wide occurrence of this phenomenon and its important role in a number of key diseases, much effort has been expended on determining key molecular factors controlling it. While factors initiating EndMT vary, a number of *in vitro* and *in vivo* studies suggested that the transforming growth factor beta (TGFβ)/bone morphogenetic protein (BMP) family growth factors are the final common pathway responsible for this endothelial cell fate change.[27,37,40] Indeed, targeted inhibition of TGFβ, BMP2, BMP4, ALK2, ALK5, endoglin, or β-glycan all result in defective heart development in mice due to the lack of EndMT.[41]

In endothelial cells, TGFβ signals via canonical and non-canonical pathways. The former occurs via the type II TGFβ receptor (TGFβR2) which activates two type I TGFβ receptors, activin like kinase (ALK) 1 and ALK5, with distinct downstream effector pathways.[42] ALK5 activation leads to phosphorylation of Smad2/3 proteins that leads to their subsequent recruitment of the common Smad4 protein and imitation of transcription of EndMT genes, perhaps in complex with the transcription factor KLF4.[43] In contrast, activation of ALK1 results in phosphorylation of Smad1/5/8 leading to increased cell proliferation and inhibition of EndMT.[44]

During embryonic heart development EndMT is mainly controlled by the TGFβ2 isoform.[45] A study using knock-out mice of all three TGFβ isoforms, $TGF\beta1^{-/-}$, $TGF\beta2^{-/-}$ and $TGF\beta3^{-/-}$, demonstrated that only TGFβ2 is essential for proper induction and cessation of embryonic cardiac EndMT.[46] However, the effects of the TGFβ2 isoform in pathological EndMT have never been investigated.

Besides canonical TGFβ signaling, TGFβ can induce EndMT through activation of ERK1/2, PI3K and p38 MAPK signaling pathways. Notch and Wnt is known to promote TGFβ-induced EndMT through modulation of the balance between different TGFβ/Smad signaling pathways (Fig. 2).

4.1. *FGF-TGFβ signaling cross-talk*

Studies evaluating FGF signaling in cultured endothelial cells demonstrated that the loss of FGF signaling, induced by a knockdown of

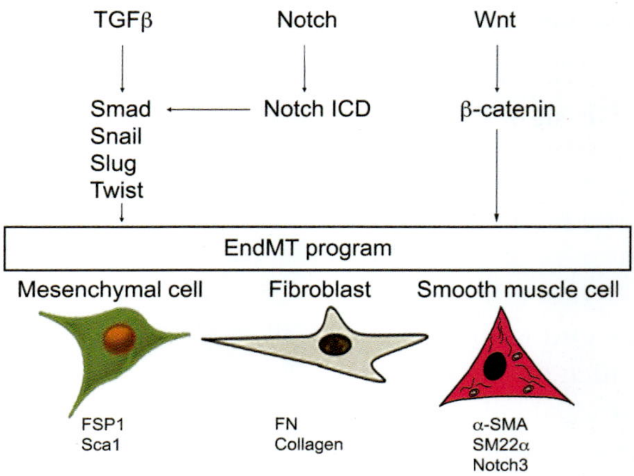

Fig. 2. Signaling pathways involved in endothelial cell to mesenchymal transition (EndMT). TGFβ, Notch, and Wnt signaling are important inducers for EndMT. Transcription factors, such as Smad, Snail, Slug, Twist, and β-catenin proteins, are important for EndMT processes. TGFβ activates Smad2 and induces Snail, Slug, and Twist expression. Smad, Snail, Slug, and Twist are then translocate to the nucleus to regulate TGFβ target gene expression and upregulate mesenchymal/fibroblast/smooth muscle markers. TGFβ also cooperates with Notch signaling through Notch ICD/Smad2 to regulate EndMT. Wnt signaling regulates EndMT through β-catenin.

either FGFR1 or a pan-FGFR adaptor protein FRS2α leads to expression of a number of smooth muscle (α-SMA, SM22α, SM-calponin, and Notch3) and mesenchymal marker (fibronectin, vimentin) genes.[16,27]

Both FGFR1 or FRS2α knockouts lead to a dramatic upregulation of a number of TGFβ signaling cascade genes including TGFβ ligands (TGFβ1, TGFβ2), receptors (TGFβR1 and TGFβR2), and a canonical downstream transcription factor Smad2. As a result, there is a profound activation of TGFβ signaling leading to expression of a number of TGFβ-dependent genes, including collagen, and a change in cell shape with endothelial cells acquiring a "mesenchymal" appearance.

Importantly, inhibition of TGFβ signaling in these setting by means of suppression or knockdown of TGFβR1 expression or kinase activity fully prevents EndMT developing demonstrating the key role played by TGFβ activation in this fate change phenomenon.[27]

4.2. *FGF regulates TGFβ activity via let-7 miRNA*

As already mentioned, activation of TGFβ signaling upon withdrawal of FGF signaling input is largely driven by an increase in TGFβR1, TGFβR2 and TGFβ2 expression levels.[27] This has been traced to an increase in mRNA half-life of these genes that, in turn, is regulated by the *let-7* family of miRNAs. The *let-7* family was originally identified in *Caenorhabditis elegans* as a regulator of developmental timing and cell proliferation.[47] The family is highly conserved across the entire phylogenetic tree and in mice and humans it consists of 11 very closely related genes located on three different chromosomes.[48]

The *let-7* miRNAs have been indirectly linked to regulation of several TGFβ family members including TGFβR1.[49] *In silico* studies show the presence of *let-7* binding sites in a number of TGFβ-related genes including TGFβR1.[27] An inhibition of endothelial FGF signaling input results in 20- to 120-fold reduction in expression of all *let-7* miRNA family members and a very significant prolongation of TGFβR1 mRNA half-life.[27]

The link between the loss of FGF signaling input, a decline in *let-7* miRNAs expression and enhanced expression of TGFβR1 is demonstrated by re-introduction of *let-7*b or *let-7*c into endothelial cells after FGFR1 or FRS2α knockdown. This fully prevents the increase in TGFβ-related gene expression and induction of EndMT program.[16,27]

4.3. *Inflammation-driven suppression of FGFR1 expression and FGF signaling in mouse transplant rejection model*

Given the importance of FGFR1 expression in regulation of EndMT, it is critical to discover factor that can affect it. Surprisingly, a trio of inflammatory mediators, IFN-γ, TNF-α, and IL-1β have all been shown to downregulate FGFR1 expression and FGF2-induced p-ERK and PLCγ activation.[27] This raises the possibility that EndMT would occur in the setting of chronic inflammation in the vessel wall such as occurs in transplant rejection or atherosclerosis. The two syndromes are characterized by a gradual buildup of neointima eventually leading to cessation of blood flow and ischemic damage to downstream tissues.

The link between FGFR1 expression, Smad2 activity, and the extent of EndMT was examined in a murine transplant rejection model. Here, implantation of a vascular graft across mouse strains and/or gender, triggers an immunological response characterized by intense vascular inflammation.[50] Thus, engraftment of an aortic segment from BALB/c mice (donor) to a C57BL/6 mouse (recipient) aorta, creates an intense graft inflammation that leads to graft rejection over the next 2 weeks. Examination of the grafts 2 weeks after transplantation shows that 10% of neointimal Notch3$^+$ and α-SMA$^+$ smooth muscle cells are of endothelial origin. Furthermore, 79% of luminal endothelial cells express Notch3$^+$ and α-SMA$^+$, indicating extensive EndMT.

An endothelial-specific FGFR1 or FRS2α deletion or inhibition of *let-7* expression by *let-7* antagomirs injection, further increases the neointima thickness, the percentage of the endothelial-derived neointimal smooth muscle cells and luminal endothelium EndMT rate. At the same time, injection of *let-7* mimics prevents EndMT development and greatly reduces neointima formation.[27]

A similar phenomenon also occurs in human disease. Immunocytochemical examination of rejected heart transplant allografts demonstrated a profound reduction of FGFR1 expression in the coronary endothelium that was accompanied by upregulation of p-Smad2, and the appearance of EndMT markers.[16]

4.4. *EndMT in atherosclerosis*

Neointima formation is a common feature of atherosclerosis, a multifactorial disease that primarily affects large- and medium-sized arteries. Despite the fact that most risk factors for atherosclerosis are present at the systemic level (such as smoking, diabetes mellitus, hypercholesterolemia, hypertension), the disease preferentially develops in certain predisposed, atheroprone regions.[51] These regions are characterized by the presence of disturbed shear stress and are typically encountered at the outer walls of vascular bifurcations and at the inner wall of vascular curvatures.[52,53] Lipid driven inflammation and blood flow/shear stress have been involved in atherosclerosis disease initiation.

Recently, endothelial-to-mesenchymal transition (EndMT) have been considered a major factor in its influence on atherosclerosis progression.[28,54] Using immunofluorescence microscopy, EndMT cells were identified in human coronary atherosclerotic plaques as cells co-expression the endothelial cell marker CD31 along with smooth muscle (SM α-actin, Notch3 and SM22α) or mesenchymal markers (fibronectin and collagen).[28] The co-expression of endothelial cell and smooth muscle/mesenchymal markers has been further confirmed in mouse Apoe null high fat diet model. Furthermore, in the porcine abdominal aortic trifurcation model, EndMT cells were detected in the outer walls of trifurcation, atheroprone regions.[54]

4.5. *Inflammation and shear stress suppression of FGFR1 expression and induction of EndMT program*

Previous work have shown that shear induces EndMT in embryonic endothelial cells.[55] Exposure to low steady shear stress (LLSS) and oscillatory shear stress (OSS) resulted in an upregulation of inflammatory gene expression, EndMT-related gene and protein expression; in contrast, high steady shear stress (LSS) was protective against inflammation and EndMT.[56–58] Krüppel-like factor 2 (KLF2) and Krüppel-like factor 4 (KLF4) are important shear stress-activated transcription factors, which regulates many of the atheroprotective effects of LSS.[59,60]

Both inflammatory cytokines and oscillatory shear stress can reduce endothelial FGFR1 expression, thereby activating TGFβ signaling.[28] *In vivo*, in the setting of high-fat diet, atherosclerosis-prone areas such as a the lesser curvature of the arch and arterial bifurcations that are characterized by low non-uniform LSS, show dramatically downregulation of FGFR1, significantly upregulation of p-Smad2 activity, and pronounced neointimal lesions. In contrast, athero-resistant areas, characterized by the absence of neointimal lesions, exhibit high levels of endothelial FGFR1 expression and low p-Smad2 activity. Inhibition of endothelial FGF signaling, achieved by a knockout of a pan-FGFR adaptor protein Frs2α, leads to earlier development of atherosclerosis and a much greater plaque burden.[28]

Their results demonstrate that EndMT underlies atherosclerosis by contributing to neointima formation, and that shear stress and inflammation modulates this process.

References

1. Powers, C.J., McLeskey, S.W., and Wellstein, A. Fibroblast growth factors, their receptors and signaling. *Endocr Relat Cancer* **7**, 165–197 (2000).
2. Mohammadi, M., Dikic, I., Sorokin, A., Burgess, W.H., *et al.* Identification of six novel autophosphorylation sites on fibroblast growth factor receptor 1 and elucidation of their importance in receptor activation and signal transduction. *Mol Cell Biol* **16**, 977–989 (1996).
3. Larsson, H., Klint, P., Landgren, E., and Claesson-Welsh, L. Fibroblast growth factor receptor-1-mediated endothelial cell proliferation is dependent on the Src homology (SH) 2/SH3 domain-containing adaptor protein Crk. *J Biol Chem* **274**, 25726–25734 (1999).
4. Mohammadi, M., Honegger, A.M., Rotin, D., Fischer, R., *et al.* A tyrosine-phosphorylated carboxy-terminal peptide of the fibroblast growth factor receptor (Flg) is a binding site for the SH2 domain of phospholipase C-gamma 1. *Mol Cell Bio* **11**, 5068–5078 (1991).
5. Cross, M.J., Lu, L., Magnusson, P., Nyqvist, D., *et al.* The Shb adaptor protein binds to tyrosine 766 in the FGFR-1 and regulates the Ras/MEK/MAPK pathway via FRS2 phosphorylation in endothelial cells. *Mol Bio Cell* **13**, 2881–2893 (2002).
6. Ornitz, D.M., and Itoh, N. The Fibroblast Growth Factor signaling pathway. *Wiley Interdiscip Rev Dev Biol* **4**, 215–266 (2015).
7. Gotoh, N. Regulation of growth factor signaling by FRS2 family docking/scaffold adaptor proteins. *Cancer Sci* **99**, 1319–1325 (2008).
8. Gotoh, N., Laks, S., Nakashima, M., Lax, I., *et al.* FRS2 family docking proteins with overlapping roles in activation of MAP kinase have distinct spatial-temporal patterns of expression of their transcripts. *FEBS Lett* **564**, 14–18 (2004).

9. Burgar, H.R., Burns, H.D., Elsden, J.L., Lalioti, M.D., *et al.* Association of the signaling adaptor FRS2 with fibroblast growth factor receptor 1 (Fgfr1) is mediated by alternative splicing of the juxtamembrane domain. *J Biol Chem* **277**, 4018–4023 (2002).
10. Kouhara, H., Hadari, Y.R., Spivak-Kroizman, T., Schilling, J., *et al.* A lipid-anchored Grb2-binding protein that links FGF-receptor activation to the Ras/MAPK signaling pathway. *Cell* **89**, 693–702 (1997).
11. Randi, A.M., Sperone, A., Dryden, N.H., and Birdsey, G.M. Regulation of angiogenesis by ETS transcription factors. *Biochem Soc Trans* **37**, 1248–1253 (2009).
12. Nentwich, O., Dingwell, K.S., Nordheim, A., and Smith, J.C. Downstream of FGF during mesoderm formation in Xenopus: the roles of Elk-1 and Egr-1. *Dev Biol* **336**, 313–326 (2009).
13. Murakami, M., Nguyen, L.T., Hatanaka, K., Schachterle, W., *et al.* FGF-dependent regulation of VEGF receptor 2 expression in mice. *J Clin Invest* **121**, 2668–2678 (2011).
14. Peters, K.G., Marie, J., Wilson, E., Ives, H.E., *et al.* Point mutation of an FGF receptor abolishes phosphatidylinositol turnover and Ca2+ flux but not mitogenesis. *Nature* **358**, 678–681 (1992).
15. Lamothe, B., Yamada, M., Schaeper, U., Birchmeier, W., *et al.* The docking protein Gab1 is an essential component of an indirect mechanism for fibroblast growth factor stimulation of the phosphatidylinositol 3-kinase/Akt antiapoptotic pathway. *Mol Cell Biol* **24**, 5657–5666 (2004).
16. Chen, P.Y., Qin, L., Tellides, G., and Simons, M. Fibroblast growth factor receptor 1 is a key inhibitor of TGFbeta signaling in the endothelium. *Sci Signal* **7**, ra90 (2014).
17. Deng, C.X., Wynshaw-Boris, A., Shen, M.M., Daugherty, C., *et al.* Murine FGFR-1 is required for early postimplantation growth and axial organization. *Genes Dev* **8**, 3045–3057 (1994).
18. Arman, E., Haffner-Krausz, R., Chen, Y., Heath, J.K., *et al.* Targeted disruption of fibroblast growth factor (FGF) receptor 2 suggests a role for FGF signaling in pregastrulation mammalian development. *Proc Natl Acad Sci USA* **95**, 5082–5087 (1998).
19. Colvin, J.S., Bohne, B.A., Harding, G.W., McEwen. D.G., *et al.* Skeletal overgrowth and deafness in mice lacking fibroblast growth factor receptor 3. *Nature Genet* **12**, 390–397 (1996).
20. Deng, C., Wynshaw-Boris, A., Zhou, F., Kuo, A., *et al.* Fibroblast growth factor receptor 3 is a negative regulator of bone growth. *Cell* **84**, 911–921 (1996).

21. Weinstein, M., Xu, X., Ohyama, K., and Deng, C.X. FGFR-3 and FGFR-4 function cooperatively to direct alveogenesis in the murine lung. *Development* **125**(18):3615–3623 (1998).
22. Gotoh, N., Manova, K., Tanaka, S., Murohashi, M., *et al.* The docking protein FRS2alpha is an essential component of multiple fibroblast growth factor responses during early mouse development. *Mol Cell Biol* **25**, 4105–4116 (2005).
23. Murakami, M., Nguyen, L.T., Zhuang, Z.W., Moodie, K.L., *et al.* The FGF system has a key role in regulating vascular integrity. *J Clin Invest* **118**, 3355–3366 (2008).
24. Pepper, M.S., and Mandriota, S.J. Regulation of vascular endothelial growth factor receptor-2 (Flk-1) expression in vascular endothelial cells. *Exp Cell Res* **241**, 414–425 (1998).
25. Hata, Y., Rook, S.L., and Aiello, L.P. Basic fibroblast growth factor induces expression of VEGF receptor KDR through a protein kinase C and p44/p42 mitogen-activated protein kinase-dependent pathway. *Diabetes* **48**, 1145–1155 (1999).
26. Oladipupo, S.S., Smith, C., Santeford, A., Park, C., *et al.* Endothelial cell FGF signaling is required for injury response but not for vascular homeostasis. *Proc Natl Acad Sci USA* **111**, 13379–13384 (2014).
27. Chen, P.Y., Qin, L., Barnes, C., Charisse, K., *et al.* FGF regulates TGF-beta signaling and endothelial-to-mesenchymal transition via control of let-7 miRNA expression. *Cell Rep* **2**, 1684–1696 (2012).
28. Chen, P.Y., Qin, L., Baeyens, N., Li, G., *et al.* Endothelial-to-mesenchymal transition drives atherosclerosis progression. *J Clin Invest* **125**, 4514–4528 (2015).
29. Correia, A.C., Moonen, J.R., Brinker, M.G., and Krenning, G. FGF2 inhibits endothelial-mesenchymal transition through microRNA-20a-mediated repression of canonical TGF-beta signaling. *J Cell Sci* **129**, 569–579 (2016).
30. Ichise, T., Yoshida, N., and Ichise, H. FGF2-induced Ras-MAPK signalling maintains lymphatic endothelial cell identity by upregulating endothelial-cell-specific gene expression and suppressing TGFbeta signalling through Smad2. *J Cell Sci* **127**, 845–857 (2014).
31. Hatanaka, K., Simons, M., and Murakami, M. Phosphorylation of VE-cadherin controls endothelial phenotypes via p120-catenin coupling and Rac1 activation. *Am J Physiol Heart Circ Physiol* **300**, H162–172 (2011).
32. Hatanaka, K., Lanahan, A.A., Murakami, M., and Simons, M. Fibroblast growth factor signaling potentiates VE-cadherin stability at adherens junctions by regulating SHP2. *PloS One* **7**, e37600 (2012).

33. Kinsella, M.G., and Fitzharris, T.P. Origin of cushion tissue in the developing chick heart: Cinematographic recordings of in situ formation. *Science* **207**, 1359–1360 (1980).
34. Markwald, R.R., Fitzharris, T.P., and Manasek, F.J. Structural development of endocardial cushions. *Am J Anat* **148**, 85–119 (1977).
35. Zeisberg, E.M., Tarnavski, O., Zeisberg, M., Dorfman, A.L., *et al.* Endothelial-to-mesenchymal transition contributes to cardiac fibrosis. *Nat Med* **13**, 952–961 (2007).
36. Cooley, B.C., Nevado, J., Mellad, J., Yang, D., *et al.* TGF-beta signaling mediates endothelial-to-mesenchymal transition (EndMT) during vein graft remodeling. *Sci Transl Med* **6**, 227ra34 (2014).
37. Maddaluno, L., Rudini, N., Cuttano, R., Bravi, L., *et al.* EndMT contributes to the onset and progression of cerebral cavernous malformations. *Nature* **498**, 492–496 (2013).
38. Arciniegas, E., Frid, M.G., Douglas, I.S., and Stenmark, K.R. Perspectives on endothelial-to-mesenchymal transition: potential contribution to vascular remodeling in chronic pulmonary hypertension. *Am J Physiol Lung Cell Mole Physiol* **293**, L1–L8 (2007).
39. Ranchoux, B., Antigny, F., Rucker-Martin, C., Hautefort, A., *et al.* Endothelial-to-mesenchymal transition in pulmonary hypertension. *Circulation* **131**, 1006–1018 (2015).
40. Medici, D., Shore, E.M., Lounev, V.Y., Kaplan, F.S., *et al.* Conversion of vascular endothelial cells into multipotent stem-like cells. *Nat Med* **16**, 1400–1406 (2010).
41. Mercado-Pimentel, M.E. and Runyan, R.B. Multiple transforming growth factor-beta isoforms and receptors function during epithelial-mesenchymal cell transformation in the embryonic heart. *Cells Tissues Organs* **185**, 146–156 (2007).
42. Goumans, M.J., Valdimarsdottir, G., Itoh, S., Rosendahl, A., *et al.* Balancing the activation state of the endothelium via two distinct TGF-beta type I receptors. *EMBO J* **21**, 1743–1753 (2002).
43. Cuttano, R., Rudini, N., Bravi, L., Corada, M., *et al.* KLF4 is a key determinant in the development and progression of cerebral cavernous malformations. *EMBO Mol Med* **8**, 6–24 (2015).
44. Lebrin, F., Goumans, M.J., Jonker, L., Carvalho, R.L., *et al.* Endoglin promotes endothelial cell proliferation and TGF-beta/ALK1 signal transduction. *EMBO J* **23**, 4018–4028 (2004).
45. Camenisch, T.D., Molin, D.G., Person, A., Runyan, R.B., *et al.* Temporal and distinct TGFbeta ligand requirements during mouse and avian endocardial cushion morphogenesis. *Dev Biol* **248**, 170–181 (2002).

46. Azhar, M., Runyan, R.B., Gard, C., Sanford, L.P., *et al.* Ligand-specific function of transforming growth factor beta in epithelial-mesenchymal transition in heart development. *Dev Dyn* **238**, 431–442 (2009).
47. Reinhart, B.J., Slack, F.J., Basson, M., Pasquinelli, A.E., *et al.* The 21-nucleotide let-7 RNA regulates developmental timing in Caenorhabditis elegans. *Nature* **403**, 901–906 (2000).
48. Roush, S. and Slack, F.J. The let-7 family of microRNAs. *Trends Cell Biol* **18**, 505–516 (2008).
49. Tzur, G., Israel, A., Levy, A., Benjamin, H., *et al.* Comprehensive gene and microRNA expression profiling reveals a role for microRNAs in human liver development. *PloS One* **4**, e7511 (2009).
50. Dietrich, H., Hu, Y., Zou, Y., Dirnhofer, S., *et al.* Mouse model of transplant arteriosclerosis: Role of intercellular adhesion molecule-1. *Arterioscler Thromb Vasc Biol* **20**, 343–352 (2000).
51. Davies, P.F., Spaan, J.A., and Krams, R. Shear stress biology of the endothelium. *Ann Biomed Eng* **33**, 1714–1718 (2005).
52. Friedman, M.H., Bargeron, C.B., Deters, O.J., Hutchins, G.M., *et al.* Correlation between wall shear and intimal thickness at a coronary artery branch. *Atherosclerosis* **68**, 27–33 (1987).
53. Ku, D.N., Giddens, D.P., Zarins, C.K., and Glagov, S. Pulsatile flow and atherosclerosis in the human carotid bifurcation. Positive correlation between plaque location and low oscillating shear stress. *Arteriosclerosis* **5**, 293–302 (1985).
54. Moonen, J.R., Lee, E.S., Schmidt, M., Maleszewska, M., *et al.* Endothelial-to-mesenchymal transition contributes to fibro-proliferative vascular disease and is modulated by fluid shear stress. *Cardiovasc Res* **108**, 377–386 (2015).
55. Egorova, A.D., Khedoe, P.P., Goumans, M.J., Yoder, B.K., *et al.* Lack of primary cilia primes shear-induced endothelial-to-mesenchymal transition. *Circ Res* **10**, 1093–1101 (2011).
56. Chappell, D.C., Varner, S.E., Nerem, R.M., Medford, R.M., *et al.* Oscillatory shear stress stimulates adhesion molecule expression in cultured human endothelium. *Circ Res* **82**, 532–539 (1998).
57. Hsiai, T.K., Cho, S.K., Wong, P.K., Ing, M., *et al.* Monocyte recruitment to endothelial cells in response to oscillatory shear stress. *FASEB J* **17**, 1648–1657 (2003).
58. Traub, O. and Berk, B.C. Laminar shear stress: mechanisms by which endothelial cells transduce an atheroprotective force. *Arterioscler Thromb Vasc Biol* **18**, 677–685 (1998).

59. Parmar, K.M., Larman, H.B., Dai, G., Zhang, Y., Jr., *et al.* Integration of flow-dependent endothelial phenotypes by Kruppel-like factor 2. *J Clin Invest* **116**, 49–58 (2006).
60. Hamik, A., Lin, Z., Kumar, A., Balcells, M., *et al.* Kruppel-like factor 4 regulates endothelial inflammation. *J Biol Chem* **282**, 13769–13779 (2007).

Chapter 5

FGF in Cardiovascular Disease

Surovi Hazarika* and Brian H Annex*,†

Abstract

The Fibroblast growth factor (FGF) receptor-ligand system is a complex family of growth factors and receptors that play crucial roles in developmental cardiovascular biology, as well as in different cardiovascular disease states in adults. In this chapter, we will discuss the structure and signaling pathways of different ligands and receptors of the FGF family. We will review the pre-clinical and clinical evidence for the role of FGF family in developmental cardiovascular biology, adult angiogenesis and wound healing, and in different cardiovascular disease states, including coronary artery disease, lower extremity peripheral vascular disease and cerebrovascular disease.

1. FGF Family of Ligands

The Fibroblast growth factor (FGF) family is a highly complex family of growth factors and receptors that play significant role in developmental pathways, adult adaptive angiogenesis and wound healing and pathological pathways such as in cancers (see chapter 1 for

*Division of Cardiovascular Medicine, Department of Medicine, University of Virginia, PO Box 800158, Charlottesville, VA, USA.

†Corresponding author: E-mail: Bha4n@virginia.edu

details).[1,2,3] This chapter will focus on the FGF receptor-ligand family in relation to cardiovascular disease.

Most FGFs are secreted glycoproteins that are sequestered to the cell surface and extracellular matrix by heparin sulphate proteoglycans (HSPG). FGFs 19, 21 and 23 bind poorly to HSPGs and thus have the potential to have both local and endocrine effects. The FGF ligands are grouped into 6 subfamilies based on differences in sequence homology and phylogeny: FGF 1 and 2, FGF 3, 7, 10, 22, FGF 4, 5 and 6, FGF 8, 17 and 18, FGF 9.16, and 20 and FGF 19, 21 and 23.[1,2,4] FGFs have a core region that consists of 120-130 amino acids arranged in 12-antiparallel beta strands, flanked by different amino and carboxyl termini. Sequence variation of the N and C-terminal tails usually accounts for the different biologic properties of the ligands. The HSPG lies within the FGF core and is composed of β1-β2 loops and parts of the β10 and β12 lops. For FGFs 19, 21 and 23, ridges in the β loops structurally hinder HSPG binding, enabling these ligands to be released to the circulation to mediate their endocrine effects.

2. FGF Family of Receptors

The primary FGF family of receptors consists of four FGFR gene products, FGFR 1-4. These receptors contain 3 extracellular immunoglobulin domains D1-D3, a single pass transmembrane domain and a cytoplasmic tyrosine kinase domain. An acidic serine rich sequence in the linker between D1 and D2, termed the acid box, which along with the D1 domain have a role in receptor autoinhibition. The D2-D3 segment of the ectodomain is necessary and sufficient for ligand binding with specificity. Several isoforms of FGFR are produced by alternative splicing. Splicing of the D1 domain and/or the acid box region results in splice variants of FGFR1-3. Alternative splicing in the second half of the D3 domain of FGFR1-3 results in FGFR1b-IIIb and FGFR1c-IIIc isoforms, which have different FGF binding specificities. FGFRs1b-IIIb are predominantly epithelial, while FGFR1c-IIIc are predominantly mesenchymal. FGFs are specific for either epithelial or mesenchymal receptors, except for FGF1, which activated both splice isoforms (for a review, see references[5–7]).

3. FGF-FGFR Specificity and Receptor Signaling

The specificity of FGF-FGFR interaction is likely regulated by primary sequence differences between the primary 18 FGF ligands and the 7 FGFR receptors (FGFR1b, FGFR1c, FGFR2b, FGFR2c, FGFR3b, FGFR3c and FGFR4). In addition, temporal and spatial expression patterns of FGFs, FGFR and HSPGs also regulate FGF-FGFR interaction. The b isoforms of the receptors are usually expressed in epithelial tissue whereas the c isoforms are usually expressed in mesenchymal tissues. In general, ligands produced in a specific tissue activate receptors in the opposite tissue. However, FGF-1 exhibits more promiscuous ligand-receptor interactions and bind to both b and c isoforms of certain FGFRs. In addition, in pathological states such as cancer or under conditions of overexpression, changes in specificity of ligand-receptor interaction are seen.[8]

After binding of ligand and cell surface HPSGs, FGFRs dimerize. The cytoplasmic tyrosine-kinase domains then transphosphorylate for receptor activation. Downstream signaling is mediated primarily via phospholipase C (PLC γ 1) and FGFR substrate 2. Further downstream signaling occurs through the Ras-mitogen-activated protein kinase (MAPK) pathway and phosphoinositide 3-kinase-Akt signaling pathways. In addition, some FGF-FGFR complexes have been shown to translocate to nucleus and initiate nuclear signaling pathways.

4. FGF System in Cardiovascular Development

FGF mutations have been identified in several human bone and connective tissue diseases.[10–13] In reference to the cardiovascular system, FGF2 deficient mice are morphologically normal, but show impaired blood pressure regulation and decreased vascular smooth muscle tone.[14,15] In addition, mice lacking FGF2 also show impaired wound healing.[16] FGF16 null mice have early embryonic death and exhibit structural defects in the heart, indicating its role in cardiac development.[17]

FGFR-1 null embryos die during gastrulation, prior to a stage in which its role in blood vessel development can be evaluated.[18] Using

Table 1. Summary table of FGF ligands, receptors, FGF like proteins, interaction of ligands and receptors.[9]

FGF Ligands	Ligands	Properties	Corresponding FGF Receptors*
FGF-1 subfamily	FGF1	High affinity for HSPG Paracrine function	All FGFRs
	FGF2		FGFR-1c, 3c > 2c, 4
FGF-4 subfamily	FGF4 FGF5 FGF6	High affinity for HSPG Paracrine, autocrine function	FGFR-1c, 2c > 3c, 4
FGF-7 subfamily	FGF3 FGF7 FGF10 FGF22	High affinity for HSPG Paracrine, autocrine function	FGFR-2b > 1b
FGF-8 subfamily	FGF8 FGF17 FGF18	High affinity for HSPG Paracrine, autocrine function	FGFR-3c > 4 >2c > 1c > 3b
FGF-9 subfamily	FGF9 FGF16 FGF20	High affinity for HSPG Paracrine function	FGFR-3c > 2c > 1c > 3b >> 4
FGF-19 subfamily	FGF19 (orthologous to mouse FGF-15) FGF21 FGF23	Poor binding to HSPG, Endocrine function	FGFR-1c, 2c, 3c, 4
FGF Homologus Factors			
FGF-Homologous Factors	FGF11 FGF12 FGF13 FGF14	Bind heparin with high affinity	Do not activate Receptors

*The specificity of FGF-FGFR interaction is complex and can be variable. This is regulated by sequence differences between the FGF ligands and the FGFR receptors, temporal and spatial expression patterns of FGFs and FGFRs, and heparin sulphate proteoglycan (HSPG) binding.

dominant-negative FGFR-1 mutant, it has been shown that FGFR-1 is required for tumor development and tumor angiogenesis.[19] In addition, adenoviral expression of dominant negative FGFR-1 results in abnormal blood vessel development and maintenance in mouse embryos cultured *in vitro*.[20]

5. FGF System in Angiogenesis and Tissue Repair

Fibroblast growth factor was originally isolated from bovine pituitary extract as a factor which stimulated division of fibroblasts in tissue culture, thus deriving its name.[21] This specific FGF was later classified as basic FGF (FGF2). Acidic fibroblast growth factor (FGF1), initially determined as an endothelial growth factor was identified several years after isolation of FGF2.[22,23] Subsequently, identification of methods to purify heparin-binding growth factors led to identification of additional several members of the FGF family. FGF-2 is a HIF-1α regulated gene, and is upregulated under hypoxia.[24]

5.1. *FGF system in angiogenesis*

FGF signaling plays a crucial role in post-natal angiogenesis and wound healing. Endothelial cells express high levels of FGFR1-IIIc and FGFR2-IIIc as well as FGF-1 and 2 and, in certain circumstances, FGF4 and FGF8b, which mediate potent pro-angiogenic signals. FGFs stimulate the formation of new blood vessels by mediating endothelial cell proliferation, altering intracellular adhesion molecules, modulating cadherins and gap junctions, and by mediating extracellular matrix degeneration.[25] *In vitro*, FGFR1 or FGFR2 activation by FGF1, FGF2 and FGF4 and FGF8b leads to endothelial cell proliferation.[25] FGF2 upregulates urokinase type plasminogen activator (uPA) in endothelial cells[26] and cancer cells,[27] while FGF 1 and FGF4 also upregulate uPA in cancer cells.[27] In addition, FGF-2 increases metalloproteinase (MMP) production in endothelial cells.[28] FGFs also modulate regulation of uPA receptors on endothelial cells,[29] thereby allowing the localization of proteolytic activity at the leading edge of the endothelial cells, thus mediating endothelial cell migration. In addition, FGF system also mediates angiogenesis by interaction with the vascular endothelial growth factor (VEGF) system.[25,30] FGF2 has been shown to induce VEGF mRNA expression in vascular endothelial cells through autocrine and paracrine mechanisms.[31] Under conditions of hypoxia, basic FGFs upregulate expression of VEGF in vascular smooth muscle.[32]

In addition to angiogenesis, FGF system has also been shown to promote arteriogenesis[33–35] and myogenesis in skeletal muscles.[35]

5.2. *FGF system in wound healing and self-renewal*

Fibroblast growth factors mediate cell division and proliferation of several cell types of cells important for wound healing, including fibroblasts and keratinocytes. Studies with local application of FGF1, FGF2, FGF4, FGF7, or FGF10 have shown the ability of these FGFs to stimulate tissue repair.[36] In addition, reduced FGF expression/responsiveness has been correlated to wound healing disorders. In diabetic mice with impaired wound healing, mRNA levels of FGF1, FGF2 and FGF7 are reduced compared to control animals.[37] Impaired would healing noticed in aged mice is associated with lower levels of FGF2, and upon addition of FGF2 angiogenic response in these mice is reduced compared to that in controls.[38] Finally, FGF-2 null mice appear superficially indistinguishable from wild-type littermates but, when challenged with full-thickness excisional wounding, show delayed healing.[16]

FGF signaling has been shown to play a significant role for self-renewal and proliferation of primitive embryonic stem cells and for proliferation, migration and lineage commitment of more differentiated stem cells.[39,40]

6. FGF System as Biomarkers in Cardiovascular Disease

FGFs 19, 21 and 23 bind poorly to HSPGs, and thereby are released and detected in the circulation. Several forms of FGFs have been studied for their role as biomarkers in predicting prognosis from specific diseases. In the heart, the soul study showed that in patients with stable CAD, higher FGF23 was independently associated with mortality and CVD events.[41] In addition, serum FGF23 has been shown to be associated with increased mortality in cardiogenic shock in patients with impaired renal function.[42] Serum FGF23 levels are also independently associated with left ventricular mass and myocardial performance index in hemodialysis patients.[43] FGF23 is independently associated with all-cause death and incident HF in community-living older persons.[44] In addition, serum levels of FGF21 are associated with carotid atherosclerosis in humans and the incurred risk is independent of the presence and extent of established risk factors.[45] In a study by Rohovsky *et al.*, basic FGF (FGF2) was significantly elevated in the serum from patients with PAD, and in patients with rest pain and ischemic ulcers, serum FGF2 was elevated upto 30 times the normal values.[46]

7. FGF System in Coronary Artery Disease

7.1. *Preclinical studies*

Several preclinical studies paved the way for clinical trials of FGF therapy in coronary artery disease. Basic fibroblast growth factor released in a controlled manner from biodegradable gelatin hydrogel injected subepicardially into infarcted myocardium in rats improved left ventricular function, improved myocardial blood flow in the peri-infarct area and increased vascular density.[47] In early studies done in a canine model of chronic myocardial ischemia, local administration of the FGF2 protein by either repeated intracoronary or direct myocardial injections enhanced collateral perfusion to ischemic myocardium.[48,49] In the first gene transfer study of FGF, using a porcine model of stress-induced ischemia, intracoronary administration of a recombinant adenovirus expressing FGF5 resulted in improved myocardial function and regional myocardial blood flow, and evidence of angiogenesis as measured by capillary number in the ischemic zone.[50] In subsequent techniques using this mode of delivery, Ad-FGF4 was shown to improve coronary blood flow and regional wall thickening during myocardial stress induced by rapid atrial pacing.[51] Similarly, FGF2 gene therapy consistently improved arteriogenesis and echocardiographic parameters of left ventricular function in a porcine model of chronic myocardial ischemia.[52,53]

7.2. *Clinicial studies*

In one of the first phase 1 human therapeutic angiogenesis studies Shumacher *et al.* administered intra-myocardial injection of FGF1 during coronary artery bypass surgery, and reported improved collateral artery growth and capillary proliferation compared to controls.[54] In subsequent phase-1 clinical studies, implanting heparin beads containing adsorbed FGF2 over ischemic myocardium at the time of bypass surgery reduced ischemic area as measured using nuclear perfusion studies, and ameliorated ischemic symptoms.[55,56] In another study, FGF2 administration reduced the size of myocardial ischemia, improved treadmill performance and reduced frequency of angina, and parenteral administration of FGF2 was tolerated well.[57] This early findings were tested in a multicenter, Phase 2, double blinded trial, (FGF Initiating Revascularization trial: FIRST). A single intracoronary infusion of FGF2 in three different doses was compared to placebo in patients with advanced CAD. Efficacy was evaluated at 90 days and 180 days by exercise tolerance test, nuclear perfusion imaging, Seattle angina questionnaire and quality of life questionnaire. FGF2 treatment conferred some benefit in quality of life the first 3 months, but later follow-ups did not show any significant difference as the placebo group continued to show improvement.[58] A sub-group analysis from this same study showed that the benefit of FGF2 therapy was noticed most in highly symptomatic patients (baseline angina frequency score ≤40) or Canadian Cardiovascular Society score of III or IV, indicating patient selection was critical for studying FGF2 therapy. There was no significant toxicity, suggesting relative safety of FGF administration as a therapeutic approach in coronary artery disease.

Preclinical studies in porcine model using FGF5 paved the way for the Angiogenic Gene Therapy (AGENT) trial.[59] The AGENT trial was a phase I/II dose-escalating study in patients with chronic stable coronary artery disease, randomized to a single intracoronary infusion of adenoviral (ad-) FGF4 or placebo. Primary end point for efficacy was treadmill exercise tolerance test (ETT) at 4 and 12 weeks. Overall, Ad-FGF4 treated patients showed a non-significant trend towards improved exercise tolerance at 4 weeks. However, a subgroup analysis showed that ETT improvement was significant in patients treated with a higher dose and in patients with baseline ETT of less than or equal to 10 minutes.

This initial trial was followed by a small randomized, double-blinded, placebo-controlled Phase II trial of Ad-FGF4 gene therapy in patients with stable angina (AGENT 2).[60] In this study, patients with stable angina with reversible ischemia on single photon emission computed tomography (SPECT) imaging were randomized to gene therapy versus placebo. Follow-up was done using clinical follow-up and repeat SPECT imaging at 8 weeks. At 8 weeks follow-up, Ad-FGF4 injection resulted in a significant reduction of ischemic defect size compared to baseline. There were no significant adverse effect noted.[60]

These suggestions of benefit in Phase I and II trials were followed by two Phase III clinical trials, Angiogenic Gene therapy-3 and 4 (AGENT-3 and 4),[61] initiated in the US and other countries. These were double blinded, placebo controlled randomized studies a low dose Ad-FGF4 and a high dose Ad-FGF4 and placebo in a 1:1:1 randomization. Patients included were between 30–75 yrs old, with chronic stable angina despite anti-anginal medications and did not require immediate PCI or CABG, or were technically unsuitable for PCI or CABG. ETT was done at baseline and at 12-weeks follow-up. Ad-FGF4 was delivered intracoronary, using divided doses between the left and right coronary systems. Analysis of pooled data from the two parallel studies showed that there was no significant difference between placebo versus Ad-FGF4 treatment for the primary endpoint of change from baseline exercise time at 12-weeks. For secondary end-points, the CCS class angina for all patients improved over placebo for the high dose group at week 12, and this was sustained at follow-up at 1 year. This change in secondary end-point was driven mainly by the female

Table 2. Summary of clinical trials with FGF in coronary artery disease.

FGF	Formulation and Dose (ug/kg)	Route of Delivery	Study Type	Patient $n =$	Primary Endpoint	Summary of Findings	References
FGF1	Peptide 10	IM	Phase I, open label	20	DS angiography	Improved collateral artery growth and capillary proliferation	54
FGF2	Peptide 0-100	IC, single	Phase I, open label	25	ETT, angiography	Reduced size of ischemia, improved ETT	57
FGF2	Peptide 0.3-0.48	IC, single	Phase I, open label	52	SAQ, ETT, MRI	Improvement in QOL and SAQ	62
FGF2	Peptide 0,10,100	Heparin beads	Phase II, double blind, randomized	24	CCS, SPECT	Reduced size of ischemia by SPECT	55,56
FGF2	Peptide 0.3-0.48	IC, iv	Phase I, open-label	59	ETT	Reduced stress-induced ischemia	63
FGF2	Peptide 0, 0.3, 3, 30	IC, single	Phase II, double blind, randomized	337	ETT, SPECT, SAQ	Improved SAQ at 90 days, not significant at 180 days	58
FGF4 AGENT-1	Adenoviral $3.3 \times 10^{\wedge 8\text{-}9}$ vp	IC	Phase I/II; double blind, randomized	79	ETT	Improvement in ETT	59
FGF4 AGENT-2	Adenoviral $10^{\wedge 10}$ vp	IC, single	Phase I/II; double blind, randomized	52	Adenosine SPECT	Improved stress-induced myocardial perfusion defects	60
FGF4 AGENT 3 and 4	Adenoviral $10^{\wedge 9}$ and $10^{\wedge 10}$ vp	IC, single	Phase III/double blind, randomized	175	ETT	Significant improvement in ETT and angina class in female cohorts; overall no effect.	61

FGF: Fibroblast Growth Factor; **IC:** Intracoronary; **IM:** Intramyocardial; **ETT:** Exercise Tolerance Test; **DS**: Digital Subtraction; **SAQ**: Seattle Angina Questionnaire; **QOL**: Quality of life; **SPECT:** Single Photon Emission; **VP**: Viral Particles.

population, as no change was observed in the male only subgroup. In subgroup analysis, the primary end-point of ETT was also significantly different in the treatment group versus placebo in the female cohort. No significant safety concerns were identified.[61]

8. FGF System in Peripheral Arterial Disease

Administration of FGF has also been examined as an approach to promote angiogenesis and arteriogenesis in peripheral arterial disease (PAD) in both pre-clinical and clinical studies.

8.1. *Pre-clinical studies*

One of the earliest studies of exogenous FGF in therapeutic angiogenesis was done in a rabbit model of hind limb ischemia. Administration of basic FGF (FGF2) into the ischemic muscle enhanced angiogenesis and growth of collateral blood vessels.[64] Subsequent studies evaluated the role of slow released FGF2 using gelatin microspheres in rabbit and rat models of hind-limb ischemia and showed evidence of angiogenesis and enhanced collateral vessel formation.[65–67] The success of therapeutic angiogenesis with FGF in pre-clinical models of limb ischemia led to clinical trials of FGF protein and gene therapy in patients with critical limb ischemia and intermittent claudication.

8.2. *Clinical studies*

Much like in coronary artery disease, clinical trials of FGF system in peripheral arterial disease met with mixed success. The first phase I clinical trial using FGF gene therapy was published in 2002.[68] In this study, a "naked" (i.e. not in a viral vector) plasmid encoding FGF1 (NV1FGF) was administered intramuscularly to the ischemic muscle of patients with critical limb ischemia with no revascularizable options. The primary objective of this study was to evaluate the safety and tolerance of increasing and repeated (2 doses) of NV1FGF. The secondary objectives were to determine the biologic activity of

NV1FGF on hemodynamic and clinical parameters associated with improved perfusion. At 12 months follow-up, there NV1FGF was found to be well tolerated. A significant improvement in ABI was noted post treatment, and significant reduction in pain and aggregate ulcer size was noted, associated with an increased transcutaneous oxygen pressure compared with baseline pretreatment values. No serious adverse effects related to FGF1 therapy were reported.

These encouraging results lead to the first phase II clinical trial with FGF in patients with CLI. In patients with CLI with no revascularizable options, administration of plasmid containing FGF1 significantly reduced the risk of amputation, but no significant difference in ulcer healing was observed.[69] There was no significant improvement in ABIs or toe brachial indices (TBIs), thereby raising the possibility that the benefit for reduced need for amputation may have been secondary to FGF affects in the micro-vasculature. However, in the follow-up TAMARIS trial, administration of a naked plasmid encoding FGF1 in patients with CLI did not provide any significant benefit in major amputation or death.[70]

Two other clinical studies have evaluated the role of therapeutic angiogenesis in patients with intermittent claudication. In an initial phase I double blinded randomized study with atherosclerotic vascular disease patients with intermittent claudication, intra-arterial basic FGF (FGF2) was well tolerated and resting calf blood flow increased significantly from baseline in a dose-dependent response.[71] In the subsequent TRAFFIC study, intra-arterial delivery of recombinant FGF2 resulted in improved peak walking time at 90 days follow-up in the group that received single injection of FGF2, but no difference in the group that received repeat injections versus single dose; either dosing group was different versus placebo.[72] Owing to the complexities of catheter based delivery further clinical development was stopped.

9. FGF System in Stroke

FGF2 is present in the neurons and glia, in the ependymal cells of ventricles and in the vascular basement membrane of cerebral blood vessels.[73] In addition, FGFR1-4 are also expressed in several

Table 3. Summary of clinical trials of FGF in PAD.

FGF	Formulation and Dose (ug or ug/kg)	Route of Delivery	Study Type	Patient (n)/	Primary Endpoint	Summary of Findings	References
FGF1	Naked plasmid DNA 500-16000	IA	Phase I/Open-labeled	51/CLI	Calf arteriography	Improved rest pain, TCO2, ABI	68
FGF1	Naked plasmid DNA	IM	Phase II	125/CLI	Ischemic Ulcer	Reduced risk of amputation, no difference in ulcer healing	69
FGF1	Naked plasmid DNA	IM	Phase III/double blind, Randomized,	525/CLI	Major amputation or death	No signififcant benefit	70
FGF2	Peptide 0,10,30	IA, single, double	Phase I/ double blind, Randomized,	19/IC	Plethysmography	Increased resting calf blood flow	71
FGF2	Recombinant FGF-2 Single/ Double dose 30	IA	Phase II/ double blind, Randomized,	174/IC	Peak Walking Time	Significant improvent of PWT at 90 days	72

FGF: Fibroblast Growth Factor; **IM:** Intramuscular; **IC:** Intermittent Claudication; **CLI:** Criticial Limb Ischemia.

different cell types in the brain.[74] In cultured neurons, FGF2 has been shown to be protective against injury in response to different inciting factors,[75] and its expression is upregulated in ischemic brain injury, indicating a potential role in adaptation/recovery from such an injury.[76,77] A growing body of literature indicates that treatment with exogenous neurotrophic growth factors may limit the extent of injury from acute ischemic stroke. FGF2 is one of the most widely studied growth factors in these settings.

9.1. *Preclinical studies*

In different animal models of cerebral ischemia, delivery of FGF2 using intra-cisternal, intravenous or intra-arterial routes led to a significant improvement in infarct volume as well as significant improvement in neurological deficits. FGF2 acts as a systemic vasodilator in different species,[78–79] but perfusion weighted-MRI do not show any perfusion differences in the ischemic penumbra, thereby indicating that the infarct-reducing effects of FGF2 are likely mediated by cytoprotection rather than by vasomotion or neo-angiogenesis.[80]

9.2. *Clinical studies*

The first study of FGF2 in acute ischemic stroke was reported by Cuevas *et al.* in 1996. This phase 1, randomized, double blind, ascending dose study for safety and tolerance of FGF2 in acute stroke patients showed that intra-arterial delivery of FGF2 was well tolerated, without any significant adverse events.[81] Although this phase I trial was promising, a subsequent phase II trial (unfortunately full data was not published) was halted because patients who received FGF2 did worse compared to the placebo group, and a companion European trial also failed to show efficacy.

10. Challenges of FGF Therapy

Despite over 30 years of extensive studies with FGF, clinical trials with FGF system has met with limited success. One must consider that the complexity of the FGF ligand–receptor system raises significant concern regarding the ability to module a specific response by transiently altering expression of a single ligand. FGFs and FGFRs are expressed in multiple cell types and mediate a multitude of signaling pathways. In addition to the different ligands, different isoforms of ligands and different receptors, there are receptors without the tyrosine kinase domain which may negatively modulate signaling. Additional complexity arises from the specificity of ligand-receptor interactions. The FGF ligand-receptor specificity is not just based on ligand structure, but also dependent on proximity of receptors to source of ligands, FGF release from extracellular matrix by heparinases, cell surface heparin sulphate proteoglycan expression and by alternative splicing of FGF receptors, which can substantially alter ligand specificity. Based on these complex interactions, FGFs may function as autocrine or paracrine factors, and certain FGFs have also been shown to function as endocrine factors. Thus, modulation of a single FGF/FGFR gene expression to alter a single process seems incomprehensible. Perhaps incorporation of a systems biology and computational approach can help us better understand and predict the consequences of alteration of a specific FGF ligand/receptor system.

Further understanding and incorporation of knowledge from external modulators of FGF-signaling is important in future designs to assess therapeutic effects of FGFs. A soluble FGFR1 receptor that lacks the transmembrane domain can bind FGF1 and FGF2 and inhibit signaling.[82] In addition, a soluble form of FGFR4 has also been recently identified, although the functional role is not clearly known.[83] Apart from the soluble receptors, other modulators of FGF signaling include: the Sprouty family of proteins;[84] FGF-binding proteins, which are carrier proteins that activate FGFs by releasing them from the extracellular matrix;[85,86] and fibronectin-leucine-rich transmembrane protein 3(FLRT3), which facilitated FGF activity.[87] Further studies are needed to understand regulation of the FGF system by epigenetic modulators such as long non-coding RNAs, micro-RNAs, and short-interferring RNAs.

References

1. Beenken, A., and Mohammadi, M. The FGF family: biology, pathophysiology and therapy. *Nat Rev Drug Discov* **8**, 235–253 (2009).
2. Turner, N., and Grose, R. Fibroblast growth factor signalling: From development to cancer. *Nat Rev Cancer* **10**, 116–129 (2010).
3. Olsen, S.K., Garbi, M., Zampieri, N., Eliseenkova, A.V., *et al.* Fibroblast growth factor (FGF) homologous factors share structural but not functional homology with FGFs. *J Biol Chem* **278**, 34226–34236 (2003).
4. Imamura, T. Physiological functions and underlying mechanisms of fibroblast growth factor (FGF) family members: recent findings and implications for their pharmacological application. *Biol Pharm Bull* **37**, 1081–1089 (2014).
5. Eswarakumar, V.P., Lax, I., and Schlessinger, J. Cellular signaling by fibroblast growth factor receptors. *Cytokine Growth Factor Rev* **16**, 139–149 (2005).
6. Klint, P., and Claesson-Welsh, L. Signal transduction by fibroblast growth factor receptors. *Front Biosci* **4**, D165–D177 (1999).
7. Rusnati, M., and Presta, M. Fibroblast growth factors/fibroblast growth factor receptors as targets for the development of anti-angiogenesis strategies. *Curr Pharm Des* **13**, 2025–2044 (2007).
8. Grose, R., and Dickson, C. Fibroblast growth factor signaling in tumorigenesis. *Cytokine Growth Factor Rev* **16**, 179–186 (2005).
9. Zhang, X., Ibrahimi, O.A., Olsen, S.K., Umemori, H., *et al.* Receptor specificity of the fibroblast growth factor family. The complete mammalian FGF family. *J Biol Chem* **281**, 15694–15700 (2006).
10. Falardeau, J., Chung, W.C., Beenken, A., Raivio, T., *et al.* Decreased FGF8 signaling causes deficiency of gonadotropin-releasing hormone in humans and mice. *J Clin Invest* **118**, 2822–2831 (2008).
11. Gribaa, M., Younes, M., Bouyacoub, Y., Korbaa, W., *et al.* An autosomal dominant hypophosphatemic rickets phenotype in a Tunisian family caused by a new FGF23 missense mutation. *J Bone Miner Metab* **28**, 111–115 (2010).
12. Milunsky, J.M., Zhao, G., Maher, T.A., Colby, R., *et al.* LADD syndrome is caused by FGF10 mutations. *Clin Genet* **69**, 349–354 (2006).
13. Tekin, M., Hismi, B.O., Fitoz, S., Ozdag, H., *et al.* Homozygous mutations in fibroblast growth factor 3 are associated with a new form of syndromic deafness characterized by inner ear, agenesis, microtia, and microdontia. *Am J Hum Genet* **80**, 338–344 (2007).

14. Dono, R., Texido, G., Dussel, R., Ehmke, H., *et al.* Impaired cerebral cortex development and blood pressure regulation in FGF-2-deficient mice. *EMBO J* **17**, 4213–4225 (1998).
15. Zhou, M., Sutliff, R.L., Paul, R.J., Lorenz, J.N., *et al.* Fibroblast growth factor 2 control of vascular tone. *Nat Med* **4**, 201–207 (1998).
16. Ortega, S., Ittmann, M., Tsang, S.H., Ehrlich, M., *et al.* Neuronal defects and delayed wound healing in mice lacking fibroblast growth factor 2. *Proc Natl Acad Sci USA* **95**, 5672–5677 (1998).
17. Lu, S.Y., Sheikh, F., Sheppard, P.C., Fresnoza, A., *et al.* FGF-16 is required for embryonic heart development. *Biochem Biophys Res Commun* **373**, 270–274 (2008).
18. Yamaguchi, T.P., Harpal, K., Henkemeyer, M., Rossant J. FGFR-1 is required for embryonic growth and mesodermal patterning during mouse gastrulation. *Genes Dev* **8**, 3032–3044 (1994).
19. Mori, S., Tran, V., Nishikawa, K., Kaneda, T., *et al.* A dominant-negative FGF1 mutant (the R50E mutant) suppresses tumorigenesis and angiogenesis. *PLoS One* **8**, e57927 (2013).
20. Lee, S.H., Schloss, D.J., and Swain, J.L. Maintenance of vascular integrity in the embryo requires signaling through the fibroblast growth factor receptor. *J Biol Chem* **275**, 33679–33687 (2000).
21. Armelin, H.A. Pituitary extracts and steroid hormones in the control of 3T3 cell growth. *Proc Natl Acad Sci USA* **70**, 2702–2706 (1973).
22. Maciag, T., Cerundolo, J., Ilsley, S., Kelley, P.R., *et al.* An endothelial cell growth factor from bovine hypothalamus: identification and partial characterization. *Proc Natl Acad Sci USA* **76**, 5674–5678 (1979).
23. Thomas, K.A., Rios-Candelore, M., and Fitzpatrick, S. Purification and characterization of acidic fibroblast growth factor from bovine brain. *Proc Natl Acad Sci USA* **81**, 357–361 (1984).
24. Calvani, M., Rapisarda, A., Uranchimeg, B., Shoemaker, R.H., *et al.* Hypoxic induction of an HIF-1alpha-dependent bFGF autocrine loop drives angiogenesis in human endothelial cells. *Blood* **107**, 2705–2712 (2006).
25. Presta, M., Dell'Era, P., Mitola, S., Moroni, E., *et al.* Fibroblast growth factor/fibroblast growth factor receptor system in angiogenesis. *Cytokine Growth Factor Rev* **16**, 159–178 (2005).
26. Presta, M., Maier, J.A., and Ragnotti, G. The mitogenic signaling pathway but not the plasminogen activator-inducing pathway of basic fibroblast growth factor is mediated through protein kinase C in fetal bovine aortic endothelial cells. *J Cell Biol* **109**, 1877–1884 (1989).

27. Billottet, C., Elkhatib, N., Thiery, J.P., and Jouanneau J. Targets of fibroblast growth factor 1 (FGF-1) and FGF-2 signaling involved in the invasive and tumorigenic behavior of carcinoma cells. *Mol Biol Cell* **15**, 4725–4734 (2004).
28. Taraboletti, G., D'Ascenzo, S., Borsotti, P., Giavazzi, R., *et al.* Shedding of the matrix metalloproteinases MMP-2, MMP-9, and MT1-MMP as membrane vesicle-associated components by endothelial cells. *Am J Pathol* **160**, 673–680 (2002).
29. Sahni, A., Sahni, S.K., Simpson-Haidaris, P.J., and Francis, C.W. Fibrinogen binding potentiates FGF-2 but not VEGF induced expression of u-PA, u-PAR, and PAI-1 in endothelial cells. *J Thromb Haemost* **2**, 1629–1636 (2004).
30. Murakami, M., Nguyen, LT., Hatanaka, K., Schachterle, W., *et al.* FGF-dependent regulation of VEGF receptor 2 expression in mice. *J Clin Invest* **121**, 2668–2678 (2011).
31. Seghezzi, G., Patel, S., Ren, C.J., Gualandris, A., *et al.* Fibroblast growth factor-2 (FGF-2) induces vascular endothelial growth factor (VEGF) expression in the endothelial cells of forming capillaries: An autocrine mechanism contributing to angiogenesis. *J Cell Biol* **141**, 1659–1673 (1998).
32. Stavri, G.T., Zachary, I.C., Baskerville, P.A., Martin, J.F., *et al.* Basic fibroblast growth factor upregulates the expression of vascular endothelial growth factor in vascular smooth muscle cells. Synergistic interaction with hypoxia. *Circulation* **92**, 11–14 (1995).
33. Deindl, E., Hoefer, I.E., Fernandez, B., Barancik, M., *et al.* Involvement of the fibroblast growth factor system in adaptive and chemokine-induced arteriogenesis. *Circ Res* **92**, 561–568 (2003).
34. Rissanen, T.T., Markkanen, J.E., Arve, K., Rutanen, J., *et al.* Fibroblast growth factor 4 induces vascular, permeability, angiogenesis and arteriogenesis in a rabbit hindlimb ischemia model. *FASEB J* **17**, 100–102 (2003).
35. Doukas, J., Blease, K., Craig, D., Ma, C., *et al.* Delivery of FGF genes to wound repair cells enhances arteriogenesis and myogenesis in skeletal muscle. *Mol Ther* **5**, 517–527 (2002).
36. Werner, S., and Grose, R. Regulation of wound healing by growth factors and cytokines. *Physiol Rev* **83**, 835–870 (2003).
37. Werner, S., Breeden, M., Hubner, G., Greenhalgh, DG., *et al.* Induction of keratinocyte growth factor expression is reduced and delayed during wound healing in the genetically diabetic mouse. *J Invest Dermatol* **103**, 469–473 (1994).

38. Swift, M.E., Kleinman, H.K., and DiPietro, L.A. Impaired wound repair and delayed angiogenesis in aged mice. *Lab Invest.* **79**, 1479–1487 (1999).
39. Coutu, D.L., and Galipeau J. Roles of FGF signaling in stem cell self-renewal, senescence and aging. *Aging (Albany NY)* **3**, 920–933 (2011).
40. Levenstein, M.E., Ludwig, T.E., Xu, R.H., Llanas, R.A., *et al.* Basic fibroblast growth factor support of human embryonic stem cell self-renewal. *Stem Cells* **24**, 568–574 (2006).
41. Parker, B.D., Schurgers, L.J., Brandenburg, V.M., Christenson, R.H., *et al.* The associations of fibroblast growth factor 23 and uncarboxylated matrix Gla protein with mortality in coronary artery disease: the Heart and Soul Study. *Ann Intern Med* **152**, 640–648 (2010).
42. Fuernau, G., Poss, J., Denks, D., Desch, S., *et al.* Fibroblast growth factor 23 in acute myocardial infarction complicated by cardiogenic shock: A biomarker substudy of the Intraaortic Balloon Pump in Cardiogenic Shock II (IABP-SHOCK II) trial. *Crit Care* **18**, 713 (2014).
43. Kirkpantur, A., Balci, M., Gurbuz, O.A., Afsar, B., *et al.* Serum fibroblast growth factor-23 (FGF-23) levels are independently associated with left ventricular mass and myocardial performance index in maintenance haemodialysis patients. *Nephrol Dial Transplant* **26**, 1346–1354 (2011).
44. Ix, J.H., Katz, R., Kestenbaum, B.R., de Boer, I.H., *et al.* Fibroblast growth factor-23 and, death, heart, failure, and cardiovascular events in community-living individuals: CHS (Cardiovascular Health Study). *J Am Coll Cardiol* **60**, 200–207 (2012).
45. Chow, W.S., Xu, A., Woo, Y.C., Tso, A.W., *et al.* Serum fibroblast growth factor-21 levels are associated with carotid atherosclerosis independent of established cardiovascular risk factors. *Arterioscler Thromb Vasc Biol* **33**, 2454–2459 (2013).
46. Rohovsky, S., Kearney, M., Pieczek, A., Rosenfield, K., *et al.* Elevated levels of basic fibroblast growth factor in patients with limb ischemia. *Am Heart J* **132**, 1015–1019 (1996).
47. Nakajima, H., Sakakibara, Y., Tambara, K., Iwakura, A., *et al.* Therapeutic angiogenesis by the controlled release of basic fibroblast growth factor for ischemic limb and heart injury: toward safety and minimal invasiveness. *J Artif Organs* **7**, 58–61 (2004).
48. Unger, E.F., Banai, S., Shou, M., Lazarous, D.F., *et al.* Basic fibroblast growth factor enhances myocardial collateral flow in a canine model. *Am J Physiol* **266**, H1588–H1595 (1994).

49. Rajanayagam, M.A., Shou, M., Thirumurti, V., Lazarous, D.F., *et al.* Intracoronary basic fibroblast growth factor enhances myocardial collateral perfusion in dogs. *J Am Coll Cardiol* **35**, 519–526 (2000).
50. Giordano, F.J., Ping, P., McKirnan, M.D., Nozaki, S., *et al.* Intracoronary gene transfer of fibroblast growth factor-5 increases blood flow and contractile function in an ischemic region of the heart. *Nat Med* **2**, 534–539 (1996).
51. Grines, C., Rubanyi, G.M., Kleiman, N.S., Marrott, P., *et al.* Angiogenic gene therapy with adenovirus 5 fibroblast growth factor-4 (Ad5FGF-4): a new option for the treatment of coronary artery disease. *Am J Cardiol* **92**, 24N–31N (2003).
52. Horvath, K.A., Doukas, J., Lu, C.Y., Belkind, N., *et al.* Myocardial functional recovery after fibroblast growth factor 2 gene therapy as assessed by echocardiography and magnetic resonance imaging. *Ann Thorac Surg* **74**, 481–486; discussion 487 (2002).
53. Heilmann, C., von Samson, P., Schlegel, K., Attmann, T., *et al.* Comparison of protein with DNA therapy for chronic myocardial ischemia using fibroblast growth factor-2. *Eur J Cardiothorac Surg* **22**, 957–964 (2002).
54. Schumacher, B., Pecher, P., and von Specht, B.U., Stegmann T. Induction of neoangiogenesis in ischemic myocardium by human growth factors: first clinical results of a new treatment of coronary heart disease. *Circulation* **97**, 645–650 (1998).
55. Laham, R.J., Sellke, FW., Edelman, E.R., Pearlman, J.D., *et al.* Local perivascular delivery of basic fibroblast growth factor in patients undergoing coronary bypass surgery: results of a phase I, randomized, double-blind, placebo-controlled trial. *Circulation* **100**, 1865–1871 (1999).
56. Ruel, M., Laham, R.J., Parker, J.A., Post, M.J., *et al.* Long-term effects of surgical angiogenic therapy with fibroblast growth factor 2 protein. *J Thorac Cardiovasc Surg* **124**, 28–34 (2002).
57. Unger, E.F., Goncalves, L., Epstein, S.E., Chew, E.Y., *et al.* 3rd, Quyyumi AA. Effects of a single intracoronary injection of basic fibroblast growth factor in stable angina pectoris. *Am J Cardiol* **85**, 1414–1419 (2000).
58. Simons, M., Annex, B.H., Laham, R.J., Kleiman, N., *et al.* Pharmacological treatment of coronary artery disease with recombinant fibroblast growth factor-2:double-blind, randomized, controlled clinical trial. *Circulation* **105**, 788–793 (2002).

59. Grines, C.L., Watkins, M.W., Helmer, G., Penny, W., *et al.* Angiogenic Gene Therapy (AGENT) trial in patients with stable angina pectoris. *Circulation* **105**, 1291–1297 (2002).
60. Grines, C.L., Watkins, M.W., Mahmarian, J.J., Iskandrian, A.E., *et al.* A randomized, double-blind, placebo-controlled trial of Ad5FGF-4 gene therapy and its effect on myocardial perfusion in patients with stable angina. *J Am Coll Cardiol* **42**, 1339–1347 (2003).
61. Henry, T.D., Grines, C.L., Watkins, M.W., Dib, N., *et al.* Effects of Ad5FGF-4 in patients with angina: an analysis of pooled data from the AGENT-3 and AGENT-4 trials. *J Am Coll Cardiol* **50**, 1038–1046 (2007).
62. Laham, R.J., Chronos, N.A., Pike, M., Leimbach, M.E., *et al.* Intracoronary basic fibroblast growth factor (FGF-2) in patients with severe ischemic heart disease: results of a phase I open-label dose escalation study. *J Am Coll Cardiol* **36**, 2132–2139 (2000).
63. Udelson, J.E., Dilsizian, V., Laham, R.J., Chronos, N., *et al.* Therapeutic angiogenesis with recombinant fibroblast growth factor-2 improves stress and rest myocardial perfusion abnormalities in patients with severe symptomatic chronic coronary artery disease. *Circulation* **102**, 1605–1610 (2000).
64. Baffour, R., Berman, J., Garb, J.L., Rhee, S.W., *et al.* Enhanced angiogenesis and growth of collaterals by in vivo administration of recombinant basic fibroblast growth factor in a rabbit model of acute lower limb ischemia: dose-response effect of basic fibroblast growth factor. *J Vasc Surg* **16**, 181–191 (1992).
65. Fujita, M., Ishihara, M., Shimizu, M., Obara, K., *et al.* Therapeutic angiogenesis induced by controlled release of fibroblast growth factor-2 from injectable chitosan/non-anticoagulant heparin hydrogel in a rat hindlimb ischemia model. *Wound Repair Regen* **15**, 58–65 (2007).
66. Kasahara, H., Tanaka, E., Fukuyama, N., Sato, E., *et al.* Biodegradable gelatin hydrogel potentiates the angiogenic effect of fibroblast growth factor 4 plasmid in rabbit hindlimb ischemia. *J Am Coll Cardiol* **41**, 1056–1062 (2003).
67. Seko, A., Nitta, N., Sonoda, A., Ohta, S., *et al.* Vascular regeneration by repeated infusions of basic fibroblast growth factor in a rabbit model of hind-limb ischemia. *AJR Am J Roentgenol* **192**, W306–310 (2009).
68. Comerota, A.J., Throm, R.C., Miller, K.A., Henry, T., *et al.* Naked plasmid DNA encoding fibroblast growth factor type 1 for the treatment of end-stage unreconstructible lower extremity ischemia: preliminary results of a phase I trial. *J Vasc Surg* **35**, 930–936 (2002).

69. Nikol, S., Baumgartner, I., Van Belle, E., Diehm, C., *et al.* Therapeutic angiogenesis with intramuscular NV1FGF improves amputation-free survival in patients with critical limb ischemia. *Mol Ther* **16**, 972–978 (2008).
70. Belch, J., Hiatt, W.R., Baumgartner, I., Driver, I.V., *et al.* Effect of fibroblast growth factor NV1FGF on amputation and death: a randomised placebo-controlled trial of gene therapy in critical limb ischaemia. *Lancet* **377**, 1929–1937 (2011).
71. Lazarous, D.F., Unger, E.F., Epstein, S.E., Stine, A., *et al.* Basic fibroblast growth factor in patients with intermittent claudication: results of a phase I trial. *J Am Coll Cardiol* **36**, 1239–1244 (2000).
72. Lederman, R.J., Mendelsohn, F.O., Anderson, R.D., Saucedo, J.F., *et al.* Therapeutic angiogenesis with recombinant fibroblast growth factor-2 for intermittent claudication (the TRAFFIC study): a randomised trial. *Lancet* **359**, 2053–2058 (2002).
73. Cuevas, P., Gimenez-Gallego, G., Martinez-Murillo, R., and Carceller, F. Immunohistochemical localization of basic fibroblast growth factor in ependymal cells of the rat lateral and third ventricles. *Acta Anat (Basel)* **141**, 307–310 (1991).
74. Gonzalez, A.M., Berry, M., Maher, P.A., Logan, A., *et al.* A comprehensive analysis of the distribution of FGF-2 and FGFR1 in the rat brain. *Brain Res* **701**, 201–226 (1995).
75. Mattson, M.P., and Furukawa, K. Programmed cell life: Anti-apoptotic signaling and therapeutic strategies for neurodegenerative disorders. *Restor Neurol Neurosci* **9**, 191–205 (1996).
76. Finklestein, S.P., Fanning, P.J., Caday, C.G., Powell, P.P., *et al.* Increased levels of basic fibroblast growth factor (bFGF) following focal brain injury. *Restor Neurol Neurosci* **1**, 387–394 (1990).
77. Speliotes, E.K., Caday, C.G., Do, T., Weise, J., *et al.* Increased expression of basic fibroblast growth factor (bFGF) following focal cerebral infarction in the rat. *Brain Res Mol Brain Res* **39**, 31–42 (1996).
78. Cuevas, P., Carceller, F., Ortega, S., Zazo, M., *et al.* Hypotensive activity of fibroblast growth factor. *Science* **254**, 1208–1210 (1991).
79. Rosenblatt, S., Irikura, K., Caday, C.G., Finklestein, S.P., *et al.* Basic fibroblast growth factor dilates rat pial arterioles. *J Cereb Blood Flow Metab* **14**, 70–74 (1994).
80. Tatlisumak, T., Takano, K., Carano, R.A., and Fisher, M. Effect of basic fibroblast growth factor on experimental focal ischemia studied by diffusion-weighted and perfusion imaging. *Stroke.* **27**, 2292–2297; discussion 2298 (1996).

81. Group FSS. Clinical Safety Trial of Intravenous Basic Fibroblast Growth Factor in Acute Stroke. *Stroke* **29**:287 (1998).
82. Guillonneau, X., Regnier-Ricard, F., Laplace, O., Jonet, L., *et al.* Fibroblast growth factor (FGF) soluble receptor 1 acts as a natural inhibitor of FGF2 neurotrophic activity during retinal degeneration. *Mol Biol Cell* **9**, 2785–2802 (1998).
83. Takaishi, S., Sawada, M., Morita, Y., Seno, H., *et al.* Identification of a novel alternative splicing of human FGF receptor 4:soluble-form splice variant expressed in human gastrointestinal epithelial cells. *Biochem Biophys Res Commun* **267**, 658–662 (2000).
84. Hacohen, N., Kramer, S., Sutherland, D., Hiromi, Y., *et al.* sprouty encodes a novel antagonist of FGF signaling that patterns apical branching of the Drosophila airways. *Cell* **92**, 253–263 (1998).
85. Abuharbeid, S., Czubayko, F., and Aigner, A. The fibroblast growth factor-binding protein FGF-BP. *Int J Biochem Cell Biol* **38**, 1463–1468 (2006).
86. Aigner, A., Butscheid, M., Kunkel, P., Krause, E., *et al.* An FGF-binding protein (FGF-BP) exerts its biological function by parallel paracrine stimulation of tumor cell and endothelial cell proliferation through FGF-2 release. *Int J Cancer* **92**, 510–517 (2001).
87. Bottcher, R.T., Pollet, N., Delius, H., Niehrs C. The transmembrane protein XFLRT3 forms a complex with FGF receptors and promotes FGF signalling. *Nat Cell Biol* **6**, 38–44 (2004).

Chapter 6

FGF Signaling in Pulmonary Hypertension

Irinna Papangeli* and Hyung J. Chun*,†

Abstract

FGF signaling is indispensable for lung development in embryogenesis and for pulmonary vascular homeostasis in adulthood. Disruption of the FGF signaling pathway can underlie various pathologies, one of which is pulmonary arterial hypertension (PAH), a subgroup of pulmonary hypertension (PH) that affects the small pulmonary arteries. PAH is characterized by unique remodeling of the pulmonary arterioles. These vascular lesions are caused by hyperproliferative endothelial and vascular smooth muscle cells, which are at least in part driven by hyperactive FGF signaling. Studies in humans and rodents have demonstrated augmented FGF signaling in the pulmonary vascular cells of PAH lungs, while inhibition of this pathway has demonstrated rescue of experimental models. In this chapter we will summarize the current state of knowledge surrounding FGF signaling in PAH, and discuss potential future directions of targeting this signaling pathway for therapeutic purposes.

*Yale Cardiovascular Research Center, Section of Cardiovascular Medicine, Yale University School of Medicine, New Haven, CT, USA.

†Corresponding author: E-mail: hyung.chun@yale.edu

1. Pulmonary Hypertension: A Rare Vasoproliferative Disease

Vascular dysfunction underlies the pathology of numerous diseases, and when presented in the pulmonary vasculature it can play a key role in the development of pulmonary hypertension (PH), a disease that remains at large incurable and is associated with high mortality rates. Pulmonary arterial hypertension (PAH) refers to the subgroup of PH that is manifested specifically in the small pulmonary arteries. PAH is a multifactorial disease characterized by vascular remodeling of the pulmonary arterioles, including formation of plexiform and concentric lesions comprised of proliferating pulmonary artery endothelial cells (PAECs) and pulmonary artery smooth muscle cells (PASMCs). The limited existing therapies have improved outcomes, but mortality remains high (15%, 30% and 45% mortality at 1, 2 and 3 years after diagnosis, respectively).[1] Given the high rate of mortality and limited modalities of treatment, further defining the disease mechanisms and identifying novel targets of potential therapy remain of utmost importance.

The maintenance of pulmonary vascular homeostasis requires a multitude of signaling cascades, both intrinsic to PAECs and PASMCs, as well as crosstalk between these cell types. A number of these signaling cascades are already targeted by existing, FDA approved drug therapies, including those driven by endothelin, prostacyclin, and cyclic GMP/nitric oxide.[2] These approved therapies are predominantly geared towards promoting pulmonary vasodilatation, although they may also provide limited anti-proliferative effects. All of the existing therapies provide symptomatic improvement, while a subset of these therapies also provide mortality benefits.[3]

A key aspect of the pulmonary vascular changes associated with PAH is the hyperproliferation of the vascular cells, including PAECs and PASMCs. Targeting this aspect of disease remains a well suited, but inadequately tested therapeutic strategy. A multitude of preclinical studies have pursued this mechanism with remarkable benefits.[4–6] These studies are highlighted by recent demonstration that pharmacologic blockade of receptor tyrosine kinases (RTKs) can rescue multiple experimental models of PH. These studies, as well as a promising case report,[7] led up to a randomized clinical trial to test the efficacy of imatinib in PAH.[8] Although a number of secondary endpoints were reached, the patients in the treatment group had greater number of intracranial hemorrhage, for which the study was terminated prematurely.

Despite the promising preclinical and anecdotal reports surrounding imatinib, its failure in a large scale randomized clinical trial demonstrates that such a broad spectrum inhibition of RTKs may not be a suitable therapeutic strategy in PAH. However, targeting cellular proliferation does have a mechanism based therapeutic role in this disease context. Alternative strategies to target proliferation, including microRNA based therapeutics,[6,9–12] inhibition of histone deacetylases,[13,14] and others provide promising strategies. Interestingly, many of these strategies do ultimately converge upon inhibiting aberrant growth factor signaling which is critically important for excess cellular proliferation. To this end, fibroblast growth factor (FGF) represents one of these aberrantly activated growth factors. The background and the rationale for targeting this pathway in PAH therapy will be the focus of this chapter.

2. FGF during Lung Development

For a comprehensive understanding of the importance of FGF signaling in PAH, it is necessary to highlight the role of this pathway during lung development (Fig. 1). As with several other branching organs, lung morphogenesis depends on tight regulation of FGF signaling in a time and tissue specific maner.[15] Fgf1, 2, 7, 8, 9, 10 and 18 are the FGF ligands expressed throughout lung development, while Fgf10 and its receptor Fgfr2 are indispensible for lung formation.[15,16] Establishment of *Nkx2-1* expression in the ventral foregut endoderm defines the initiation of lung bud development, concomitant with the separation of the single foregut tube to dorsal (future esophagus) and ventral (future trachea) tubes.[17,18] During this initial stage of pulmonary cell fate specification, high levels of Fgf1 and Fgf2 secreted from the adjacent cardiac mesoderm promote *Nkx2-1* expression, via Fgfr1 and Fgfr4, as shown in *ex vivo* foregut explants.[19]

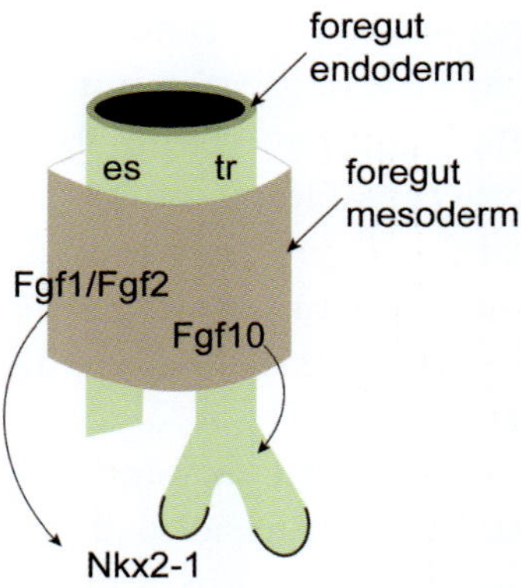

(a)

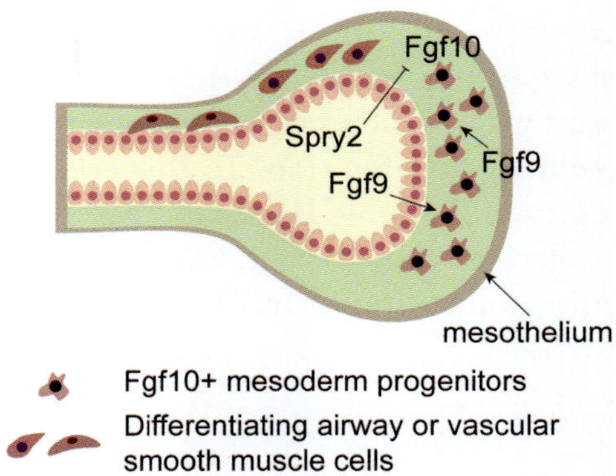

(b)

Fig. 1. FGF signaling in lung development. (a) Early patterning of the anterior foregut begins with a single tube surrounded by mesoderm. Fgf1/2 signals from the foregut mesoderm induce Nkx2-1 expression in the foregut endoderm, which establishes the native lung buds. Localized expression of Fgf10 promotes lung bud outgrowth. (b) During distal bud morphogenesis Fgf10 is required in the distal mesoderm for cell proliferation and bud outgrowth. Mesothelial and epithelial Fgf9 signaling promotes the proliferation of Fgf10-expressing mesenchymal cells. *Spry2* blocks FGF signaling at already established bud tips and prevents further bud outgrowth. Es, esophagus; Tr, trachea.

The localized expression of Fgf1, Fgf2 and Fgf10 in the ventral mesoderm adjacent to the foregut epithelium indicates the role of FGF signaling in pulmonary lineage specification along the anterior-posterior axis, contrary to BMP signals, which are required along the dorsal-ventral axis.[18] Mesodermally expressed Fgf10 signals to Fgfr2b in the endoderm, initiates lung bud outgrowth and drives branching morphogenesis.[17,18] Fgf10/Fgfr2b act through prolonged activation of the PI3K-AKT pathway, which results in cell migration and epithelial branching.[20] These Fgf10 positive cells eventually give rise to airway and smooth muscle cells, as well as lipofibroblasts.[16] Similar to Fgf10, Fgf7 is expressed in the mesenchyme and also signals through Fgfr2b, however, Fgf7 signaling leads to Fgfr2b degradation, promoting transient Fgfr2b signaling and proliferation.[16,20] Fgf10 signaling induces genes like *Bmp4* at the epithelial bud tips that inhibit lung bud outgrowth, thus allowing for branching following primary bud formation.[17,18] Branching of the lung buds is a complex process that involves multiple antagonists, such as *Spry2*, a negative regulator of receptor tyrosine kinase (RTK) signaling in the epithelium.[17,18] *Spry2* is induced by Fgf10 and is essentially utilized to block FGF signaling and prevent further bud outgrowth of cells at already established bud tips.[18] During midgestation, mesothelial Fgf9 promotes proliferation in Fgf10 expressing cells and inhibits smooth muscle cell (SMC) differentiation, thus maintaining a progenitor pool.[16] Subsequently, endodermal Bmp4 signals to the distal mesoderm and initiates SMC differentiation, a process that is in part dependent on Sonic Hedgehog (SHH) induced Fgf10 downregulation.[15,17] Epithelial Fgf9 is additionally important for branching morphogenesis through autocrine activation of epithelial FGFRs (Fgfr1c/2c) or indirect regulation of mesenchymal FGF signaling,[16] while (along with SHH) it promotes mesenchymal Vegfa expression, which is essential for pulmonary vasculature development.[18] Recent studies have shown that Fgf10 is required for establishing or maintaining tracheal basal cell density.[21] During late stages of lung development, different sets of FGF signaling molecules, including Fgf7,[22] Fgf18,[23] Fgfr3 and Fgfr4, function cooperatively to regulate processes such as alveolar formation.[24,25]

3. FGF Signaling in PAH

While FGF signaling has been found to be critical for maintenance of vascular integrity and homeostasis,[26] a number of disease states

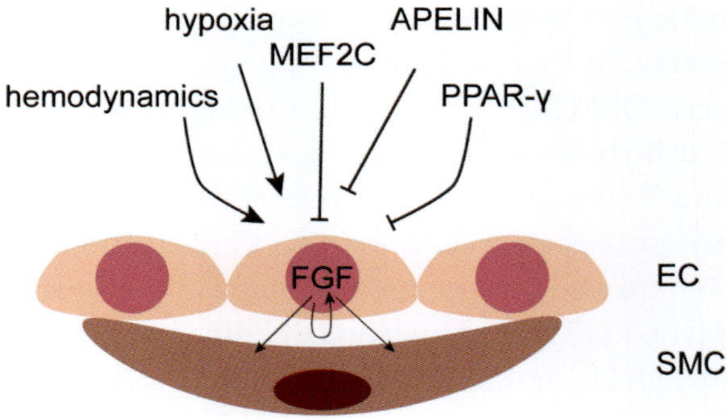

Fig. 2. Factors influencing FGF signaling in pulmonary hypertension. Secreted FGF from the endothelium induces endothelial and smooth muscle cell proliferation. Hemodynamic forces such as shear stress and stretch, as well as hypoxic conditions, induce FGF2 expression in the endothelium. Negative regulators of FGF2 include MEF2C, PPAR-γ and Apelin as well as downstream microRNAs.

have also identified aberrantly increased FGF signaling as a disease feature, many of which will be described elsewhere in this book. Such has also been found to be the case in the context of PAH. A number of mechanisms involving aberrant growth factor signaling in PAH have been investigated, including those driven by hypoxia,[27,28] variant hemodynamic forces,[29,30] cell surface receptors such as GPCRs,[6] and aberrant transcriptional machinery involving factors such as PPAR-γ[31] and MEF2 (Fig. 2).[14] With respect to FGF signaling, all of these above-mentioned factors have been found to contribute to aberrantly increased FGF2 expression, which is emerging as a key contributor to the pathogenesis of PAH. In hypoxic conditions, PAECs produce growth factors, including FGF2, that are potent mitogens for PASMCs, increasing their proliferative capacity.[27,28] Both PAECs and PASMCs have been shown to induce FGF2 promoter activity in response to cyclic stretch and shear stress, with PASMCs displaying more robust responses.[29,30] GPCR signaling mediated by the peptide ligand apelin (APLN) and its cognate receptor APJ (APLNR/AGTRL1), was shown to be critical for regulating FGF2 and FGFR1 levels in PAECs.[6] Lastly, PPAR-γ, which has been found to have impaired activity in PAH,[32] was found to inhibit FGF2 expression in PAECs.[31]

Such a role for FGF2 as a mediator of PAH pathogenesis has been demonstrated in both the preclinical and clinical contexts. One of the initial studies implicating FGF2 in PAH originated from Judah Folkman and colleagues, where both the serum and urine levels of FGF2 were found to be markedly elevated in PAH subjects compared to controls.[33] Although this initial study did not identify the specific cellular subtypes that express FGF2, subsequent studies demonstrated that the PAECs from PAH subjects expressed significantly higher levels of FGF2 compared to control PAECs.[34] This work, as well as subsequent work, demonstrated that this aberrant expression of FGF2 promotes aberrant proliferation of both PAECs and PASMCs.[34,35] Moreover, these studies found that abrogating this increase in FGF2 expression via an RNA interference approach led to rescue of experimental models of PH.[34] Overall, this body of data provide a compelling combination of both experimental and clinical support for the role of aberrant FGF signaling as a key mediator of PAH.

4. Endothelial Derived FGF2 in PAH

As discussed above, FGF2 expression was found to be significantly increased in PAH PAECs.[34] Multiple factors may be contributing to this phenomenon, but recent work from our laboratory identified a key mechanism by which microRNA mediated regulation of FGF signaling may be a key mediator of its regulation in the pulmonary vasculature.[6] PAECs from PAH subjects also produced more FGFR1, suggesting they have enhanced response to FGF2 signaling.[35] The increased autocrine FGF2 signaling observed in PAH PAECs confers a highly proliferative EC phenotype through enhanced activity of ERK1 and ERK2 (ERK1/2), concomitant with decreased apoptosis signaling via elevated BCL2 and BCL-xL antiapoptotic factors.[35] We also identified a key upstream regulatory mechanism involving APLN, where decreased APLN expression levels in the serum and PAECs of subjects with PAH[6,36–38] resulted in increased FGF2 and FGFR1 expression via decreased expression of two endothelial microRNAs (miRs), namely miR-424 and miR-503.[6] In both human PAECs and mouse lung endothelial cells, decreased APLN correlated with increased FGF2 and FGFR1 levels, and conversely APLN overexpression induced decrease in FGF2 and FGFR1 levels.[6] This interaction was mediated by direct binding of miR-424 and miR-503 to the 3′ untranslated regions (UTRs) of FGF2 and FGFR1, which led to decreased expression levels of these genes.[6] As a consequence, reduced ERK1/2 phosphorylation, which is downstream of FGF2/FGFR1 signaling, was observed in cells overexpressing miR-424 and miR-503.[6] Interestingly, reduced levels of miR-424 and miR-503, correlating with increased FGF2 expression, were found in PAH PAECs compared to PAECs from control subjects.[6] Restoration of miR-424 and miR-503 expression induced cell cycle arrest and had anti-proliferative effects in PAECs.[6] Intranasal delivery of lentiviral constructs encoding for miR-424 and miR-503 rescued PH in two different experimental rat models, as demonstrated by: (1) decrease in right ventricular systolic pressure (RVSP), (2) decrease in right ventricle to left ventricle plus septum weight ratio, and (3) reduction in number of muscularized or obliterated pulmonary microvessels.[6] Importantly, we found significantly decreased expression of FGF2 and FGFR1 in rats administered miR-424 and miR-503, further supporting the mechanistic role of FGF2 in PH pathogenesis.

Additional studies in ovine PAECs have shown that PPAR-γ is required to inhibit FGF2 and other angiogenic growth factor expression.[31] Pharmacological inhibition of PPAR-γ was found to be upregulate FGF2 at both the mRNA and protein levels.[31] Loss of PPAR-γ was found to be associated with greater number of cells remaining in S phase via upregulation of Cyclin C, which promotes exit at G_0/G_1 and reentry into S phase in cell cycle.[31] FGF2 was also found elevated in the lung tissue of an experimental model of PH in lambs, which was associated with low levels of PPAR-γ.[29,31]

5. FGF Signaling in Pulmonary Arterial Smooth Muscle Cells

Previous studies demonstrated that conditioned media derived from PAH PAECs can induce greater proliferation of PASMCs in an FGF2 dependent manner.[6,34] It has also been shown that exogenous FGF signaling affects PASMC proliferation,[39–41] migration,[41] and matricellular protein expression,[42] as well as pulmonary pericyte proliferation and migration.[43] Proliferation of PASMCs in response to FGF2 is mediated in part through ERK1/2, which leads to nuclear translocation of NF-κB.[39] NF-κB subsequently activates monocyte chemoattractant protein 1 (MCP-1), and plasminogen activator inhibitor 1 (PAI-1),[39] factors with known roles in vascular remodeling by promoting monocyte activation and migration,[44] fibrin deposition[45] and PASMC proliferation.[44,46] Inhibiting NF-κB using IMD-0354, a selective chemical inhibitor, was sufficient to improve PH in terms of RVSP, muscularization and medial hypertrophy of pulmonary arterioles.[39] Interestingly, IMD-0354 reduced the total FGF2 mRNA levels in whole lung lysates of monocrotaline induced PH rats,[39] suggesting it may have a primary effect on endothelial FGF signaling, considering that the main source of FGF2 in PAH appears to be the pulmonary endothelium.[34]

In addition to FGF2, FGF1 has also been found to be a potent mitogen for PASMCs. In response to FGF1, PASMCs upregulate the expression of matricellular proteins periostin (PN) and osteopontin (OPN).[42] Both proteins are known to actively participate in ECM remodeling in disease models and were found increased in lungs of hypoxia induced PH rat models.[42] Upregulation of PN and OPN is considered part of a generalized ECM response to hypoxic stress that likely contributes to the progression of PH.[42] The PN induction is mediated through the PI3K/p70S6K, Ras/MEK1/2 and Ras/p38MAPK pathways whereas the OPN induction is through the Ras/MEK1/2 and Ras/JNK pathways.[42] This FGF dependent response of matricellular proteins in PASMCs demonstrates the diverse mechanisms involved in vascular remodeling in PH.[42]

Another important modifier of PAH that is indirectly associated with FGF signaling is endothelin 1 (ET-1), a potent vasoconstrictor expressed by endothelial cells.[47] Both ET-1 and its subtype A receptor (ET-AR) are increased in the lungs of hypoxia induced PH rat or mouse models,[48] while ET-1 plasma levels are elevated in patients with some forms of PH.[49] This ligand-receptor pair is an important mediator of hypoxic pulmonary vasoconstriction and vascular remodelling.[48] In PASMCs, hypoxia does not affect expression of

ET-AR directly; however, hypoxia induced FGF1 and FGF2 expression increased ET-AR levels in a time and dose dependent manner.[48] This FGF signalling is mediated through FGFR1/MAP kinase downstream pathways[48] and potentiates ET-1/ET-AR signalling towards the progression of PH.[48]

Similar to ET-1, Atrial Natriuretic Peptide (ANP) is involved in pulmonary vasoconstriction and vascular remodeling.[50] It is a selective relaxant of pulmonary arteries that counteracts the pathologic mechanisms of PH and as such, is found elevated in the plasma of hypoxia induced PH rat or mouse models as a compensatory mechanism.[50] Homozygous deletion of ANP in mice leads to severe PH and RV hypertrophy in response to hypoxia.[50] In PASMCs, FGF1 and FGF2 can induce a rapid reduction of NPR-C, the ANP receptor that lacks the intracellular signaling domain found in NPR-A and B receptors and is responsible for ligand clearance.[51] This most likely represents a developmental adaptation that enhances the ANP mediated vasodilatory and antiproliferative effects in the hypoxic lung.[50]

Lastly, pericytes, a recently identified cell population that contributes to the progression of the disease are also shown to be FGF dependent.[43] FGF2 is demonstrated to induce pericyte migration, and through Cyclin D1 pericyte proliferation.[43]

6. Current FGF based Clinical Approaches

Several promising studies in animal models that focus on general RTK or selective FGF pathway inhibition have set the basis for future clinical studies in PAH. Less selective RTK inhibitors such as imatinib,[8] sorafinib,[5] and suramin[52] have shown evidence of efficacy in clinical and preclinical studies in PAH. However, another member of this class of drugs, namely dasatinib, was actually found to be associated with increased incidence of PAH.[53] Although RTKs may be a potentially promising target in PAH, greater degree of selectivity is likely critical for this class of drugs to be further considered in PAH based on these results.

To date, no study has selectively targeted FGF signaling as a therapeutic strategy in PAH. However, such therapeutic strategy is

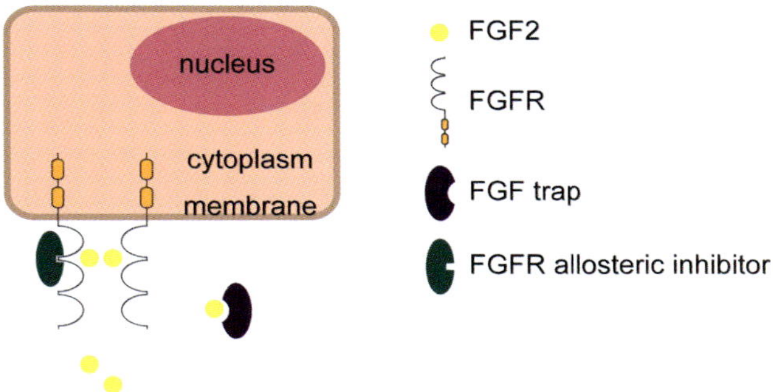

Fig. 3. Current pharmacological approaches targeting FGF signaling. Two different strategies are being used to target FGF signaling. FGF decoy receptors, such as FP-1039, selectively target the mitogenic FGF ligands. Allosteric FGF inhibitors (SSR128129E) can efficiently reduce downstream FGF signaling, without affecting ligand binding domains.

currently being developed, primarily in the cancer field (Fig. 3). A soluble decoy receptor, FP-1039 (also known as GSK3052230), was recently developed to selectively target the mitogenic FGF ligands.[54] FP-1039 was shown to be highly effective in reducing tumor growth of cancer xenograft models with genetic aberrations in the FGF pathway in mice, without affecting serum calcium and phosphate levels, which are regulated by FGF23.[54] *In vitro* it was shown to inhibit both tumor cell proliferation and angiogenesis.[54] A different approach has been to target allosteric sites of RTKs (domains that are topographically different from these used by orthosteric ligands), rather than prevent ligand binding or target kinase inhibition through small molecules.[55] SSR (SSR128129E) was identified as a selective, multi-FGFR inhibitor that can affect FGF related EC proliferation, migration, survival and tube formation *in vitro*, by binding to the extracellular domain of FGFRs and inhibiting downstream signaling related to receptor internalization.[55] Interestingly, SSR inhibited FGF2 induced ERK1/2 phosphorylation and intracellular accumulation of early endosomal antigen 1 (EEA1) positive vesicles, which are the destination of internalized FGFRs.[55] *In vivo*, oral administration of SSR improved the clinical symptoms of arthritis in a mouse model, while it reduced primary tumor growth and metastasis in several xenograft mouse models of FGFR and FGF expressing cancer cell lines.[55]

The availability and ongoing pursuit of these FGF selective agents provide key tools to pursue FGF inhibition as a therapeutic strategy in PAH. At least one of these agents is already in a Phase 1 clinical trial investigating its anticancer efficacy in tumors with abnormal dependence on FGF pathway signaling.

7. Limitations and Perspectives

Although excess FGF signaling may be a key mechanistic driver in PAH, broad inhibition of FGF signaling may pose a challenge due to its effects on other vascular beds, which may depend on FGF to maintain a homeostatic balance. Moreover, further refinements in therapeutic strategy, including the optimal duration and route of therapy, will need to be defined specifically for the PAH context. Nevertheless, given that ongoing clinical trials are emerging to show feasibility and tolerability of anti-FGF therapy in humans, clinical trials in PAH targeting FGF may not be too far from reality.

References

1. Humbert, M., *et al.* Survival in patients with idiopathic, familial, and anorexigen-associated pulmonary arterial hypertension in the modern management era. *Circulation* **122**, 156–163 (2010).
2. Galie, N. *et al.* Guidelines for the diagnosis and treatment of pulmonary hypertension. *Eur Respir J* **34**, 1219–1263 (2009).
3. Barst, R.J., *et al.* A comparison of continuous intravenous epoprostenol (prostacyclin) with conventional therapy for primary pulmonary hypertension. *N Engl J Med* **334**, 296–301 (1996).
4. Schermuly, R.T., *et al.* Reversal of experimental pulmonary hypertension by PDGF inhibition. *J Clin Invest* **115**, 2811–2821 (2005).
5. Moreno-Vinasco, L., *et al.* Genomic assessment of a multikinase inhibitor, sorafenib, in a rodent model of pulmonary hypertension. *Physiol Genomics* **33**, 278–291 (2008).
6. Kim, J., *et al.* An endothelial apelin-FGF link mediated by miR-424 and miR-503 is disrupted in pulmonary arterial hypertension. *Nat Med* **19**, 74–82 (2013).

7. Ghofrani, H.A., Seeger, W. and Grimminger, F. Imatinib for the treatment of pulmonary arterial hypertension. *N Engl J Med* **353**, 1412–1413 (2005).
8. Hoeper, M.M., *et al.* Imatinib mesylate as add-on therapy for pulmonary arterial hypertension: results of the randomized IMPRES study. *Circulation* **127**, 1128–1138 (2013).
9. Bertero, T., *et al.* Systems-level regulation of microRNA networks by miR-130/301 promotes pulmonary hypertension. *J Clin Invest* **124**, 3514–3528 (2014).
10. Pohl, N.M., Fernandez, R.A., Smith, K.A. and Yuan, J.X. Deacetylation of MicroRNA-124 in fibroblasts: role in pulmonary hypertension. *Circ Res* **114**, 5–8 (2014).
11. Parikh, V.N., *et al.* MicroRNA-21 integrates pathogenic signaling to control pulmonary hypertension: Results of a network bioinformatics approach. *Circulation* **125**, 1520–1532 (2012).
12. Pullamsetti, S.S., *et al.* Inhibition of microRNA-17 improves lung and heart function in experimental pulmonary hypertension. *Am J Respir Crit Care Med* **185**, 409–419 (2012).
13. Zhao, L., *et al.* Histone deacetylation inhibition in pulmonary hypertension: therapeutic potential of valproic acid and suberoylanilide hydroxamic acid. *Circulation* **126**, 455–467 (2012).
14. Kim, J., *et al.* Restoration of impaired endothelial myocyte enhancer factor 2 function rescues pulmonary arterial hypertension. *Circulation* **131**, 190–199 (2015).
15. Moura, R.S., Coutinho-Borges, J.P., Pacheco, A.P., Damota, P.O., *et al.* FGF signaling pathway in the developing chick lung: expression and inhibition studies. *PLoS One* **6**, e17660 (2011).
16. Volckaert, T. & De Langhe, S.P. Wnt and FGF mediated epithelial-mesenchymal crosstalk during lung development. *Dev Dyn* **244**, 342–366 (2015).
17. Morrisey, E.E. and Hogan, B.L. Preparing for the first breath: genetic and cellular mechanisms in lung development. *Dev Cell* **18**, 8–23 (2010).
18. Hines, E.A. and Sun, X. Tissue crosstalk in lung development. *J Cell Biochem* **115**, 1469–1477 (2014).
19. Serls, A.E., Doherty, S., Parvatiyar, P., Wells, J.M. *et al.* Different thresholds of fibroblast growth factors pattern the ventral foregut into liver and lung. *Development* **132**, 35–47 (2005).
20. Francavilla, C., *et al.* Functional proteomics defines the molecular switch underlying FGF receptor trafficking and cellular outputs. *Mol Cell* **51**, 707–722 (2013).

21. Volckaert, T., *et al.* Localized FGF10 expression is not required for lung branching morphogenesis but prevents differentiation of epithelial progenitors. *Development* **140**, 3731–3742 (2013).
22. Padela, S., *et al.* A critical role for fibroblast growth factor-7 during early alveolar formation in the neonatal rat. *Pediatr Res* **63**, 232–238 (2008).
23. Franco-Montoya, M.L., *et al.* Profiling target genes of FGF18 in the postnatal mouse lung: possible relevance for alveolar development. *Physiol Genomics* **43**, 1226–1240 (2011).
24. Weinstein, M., Xu, X., Ohyama, K. and Deng, C.X. FGFR-3 and FGFR-4 function cooperatively to direct alveogenesis in the murine lung. *Development* **125**, 3615–3623 (1998).
25. Srisuma, S., *et al.* Fibroblast growth factor receptors control epithelial-mesenchymal interactions necessary for alveolar elastogenesis. *Am J Respir Crit Care Med* **181**, 838–850 (2010).
26. Murakami, M., *et al.* FGF-dependent regulation of VEGF receptor 2 expression in mice. *J Clin Invest* **121**, 2668–2678 (2011).
27. Vender, R.L. Role of endothelial cells in the proliferative response of cultured pulmonary vascular smooth muscle cells to reduced oxygen tension. *In Vitro Cell Dev Biol* **28A**, 403–409 (1992).
28. Vender, R.L., Clemmons, D.R., Kwock, L. and Friedman, M. Reduced oxygen tension induces pulmonary endothelium to release a pulmonary smooth muscle cell mitogen(s). *Am Rev Respir Dis* **135**, 622–627 (1987).
29. Wedgwood, S., *et al.* Fibroblast growth factor-2 expression is altered in lambs with increased pulmonary blood flow and pulmonary hypertension. *Pediatr Res* **61**, 32–36 (2007).
30. Quinn, T.P., Schlueter, M., Soifer, S.J. and Gutierrez, J.A. Cyclic mechanical stretch induces VEGF and FGF-2 expression in pulmonary vascular smooth muscle cells. *Am J Physiol Lung Cell Mol Physiol* **282**, L897–L903 (2002).
31. Tian, J., *et al.* Effect of PPARgamma inhibition on pulmonary endothelial cell gene expression: gene profiling in pulmonary hypertension. *Physiol Genomics* **40**, 48–60 (2009).
32. Gong, K., *et al.* Hypoxia induces downregulation of PPAR-gamma in isolated pulmonary arterial smooth muscle cells and in rat lung via transforming growth factor-beta signaling. *Am J Physiol Lung Cell Mol Physiol* **301**, L899–L907 (2011).

33. Benisty, J.I., *et al.* Elevated basic fibroblast growth factor levels in patients with pulmonary arterial hypertension. *Chest* **126**, 1255–1261 (2004).
34. Izikki, M., *et al.* Endothelial-derived FGF2 contributes to the progression of pulmonary hypertension in humans and rodents. *J Clin Invest* **119**, 512–523 (2009).
35. Tu, L., *et al.* Autocrine fibroblast growth factor-2 signaling contributes to altered endothelial phenotype in pulmonary hypertension. *Am J Respir Cell Mol Biol* **45**, 311–322 (2011).
36. Chandra, S.M., *et al.* Disruption of the apelin-APJ system worsens hypoxia-induced pulmonary hypertension. *Arterioscler Thromb Vasc Biol* **31**, 814–820 (2011).
37. Alastalo, T.P., *et al.* Disruption of PPARgamma/beta-catenin-mediated regulation of apelin impairs BMP-induced mouse and human pulmonary arterial EC survival. *J Clin Invest* **121**, 3735–3746 (2011).
38. Goetze, J.P., *et al.* Apelin: A new plasma marker of cardiopulmonary disease. *Regul Pept* **133**, 134–138 (2006).
39. Hosokawa, S., *et al.* Pathophysiological roles of nuclear factor kappaB (NF-kB) in pulmonary arterial hypertension: effects of synthetic selective NF-kB inhibitor IMD-0354. *Cardiovasc Res* **99**, 35–43 (2013).
40. Schultz, K., Fanburg, B.L. and Beasley, D. Hypoxia and hypoxia-inducible factor-1alpha promote growth factor-induced proliferation of human vascular smooth muscle cells. *Am J Physiol Heart Circ Physiol* **290**, H2528–H2534 (2006).
41. Tu, L., *et al.* A critical role for p130Cas in the progression of pulmonary hypertension in humans and rodents. *Am J Respir Crit Care Med* **186**, 666–676 (2012).
42. Li, P., Oparil, S., Feng, W. and Chen, Y.F. Hypoxia-responsive growth factors upregulate periostin and osteopontin expression via distinct signaling pathways in rat pulmonary arterial smooth muscle cells. *J Appl Physiol (1985)* **97**, 1550–1558; discussion 1549 (2004).
43. Ricard, N., *et al.* Increased pericyte coverage mediated by endothelial-derived fibroblast growth factor-2 and interleukin-6 is a source of smooth muscle-like cells in pulmonary hypertension. *Circulation* **129**, 1586–1597 (2014).
44. Sanchez, O., *et al.* Role of endothelium-derived CC chemokine ligand 2 in idiopathic pulmonary arterial hypertension. *Am J Respir Crit Care Med* **176**, 1041–1047 (2007).

45. Pinsky, D.J., *et al.* Coordinated induction of plasminogen activator inhibitor-1 (PAI-1) and inhibition of plasminogen activator gene expression by hypoxia promotes pulmonary vascular fibrin deposition. *J Clin Invest* **102**, 919–928 (1998).
46. Diebold, I., Djordjevic, T., Hess, J. and Gorlach, A. Rac-1 promotes pulmonary artery smooth muscle cell proliferation by upregulation of plasminogen activator inhibitor-1: Role of NFkappaB-dependent hypoxia-inducible factor-1alpha transcription. *Thromb Haemost* **100**, 1021–1028 (2008).
47. Chen, Y.F. and Oparil, S. Endothelin and pulmonary hypertension. *J Cardiovasc Pharmacol* **35**, S49–S53 (2000).
48. Li, P., Oparil, S., Sun, J.Z., Thompson, J.A. *et al.* Fibroblast growth factor mediates hypoxia-induced endothelin — a receptor expression in lung artery smooth muscle cells. *J Appl Physiol* (1985) **95**, 643–651; discussion 863 (2003).
49. Stewart, D.J., Levy, R.D., Cernacek, P. and Langleben, D. Increased plasma endothelin-1 in pulmonary hypertension: marker or mediator of disease? *Ann Intern Med* **114**, 464–469 (1991).
50. Chen, Y.F. Atrial natriuretic peptide in hypoxia. *Peptides* **26**, 1068–1077 (2005).
51. Sun, J.Z., Oparil, S., Lucchesi, P., Thompson, J.A. *et al.* Tyrosine kinase receptor activation inhibits NPR-C in lung arterial smooth muscle cells. *Am J Physiol Lung Cell Mol Physiol* **281**, L155–L163 (2001).
52. Izikki, M., *et al.* The beneficial effect of suramin on monocrotaline-induced pulmonary hypertension in rats. *PLoS One.* **8**, e77073 (2013).
53. Montani, D., *et al.* Pulmonary arterial hypertension in patients treated by dasatinib. *Circulation* **125**, 2128–1237 (2012).
54. Harding, T.C., *et al.* Blockade of nonhormonal fibroblast growth factors by FP-1039 inhibits growth of multiple types of cancer. *Sci Transl Med* **5**, 178ra39 (2013).
55. Herbert, C., *et al.* Molecular mechanism of SSR128129E, an extracellularly acting, small-molecule, allosteric inhibitor of FGF receptor signaling. *Cancer Cell* **23**, 489–501 (2013).

Chapter 7

Therapeutic Potential of Allosteric Modulation of FGF Receptors

Frederik De Smet*,† and Peter Carmeliet‡,§

Abstract

Fibroblast Growth Factor receptors (FGFR) function as receptor tyrosine kinases (RTKs) and represent major targets for drug development. Traditional RTK inhibitors block orthosteric binding of ligands and substrates. Allosteric ligands provide a rich source of possible drug targets with clear therapeutic advantages, but the complexity of their mechanisms makes the discovery and development of allosteric drugs challenging. The FGFR has recently been targeted by SSR128129E (SSR), an extracellularly-acting small-molecule allosteric modulator. In this chapter, we discuss the mode-of-action of SSR and the advantages and challenges associated with allosteric targeting of FGFRs. The allosteric

*The Switch Laboratory, Department of Cellular and Molecular Medicine, University of Leuven, Leuven, B-3000, Belgium.

†The Switch Laboratory, Flanders Institute for Biotechnology (VIB), Leuven, B-3000, Belgium.

‡Laboratory of Angiogenesis and Vascular Metabolism, Department of Oncology, KU Leuven, Leuven, B-3000, Belgium.

§Laboratory of Angiogenesis and Vascular Metabolism, Vesalius Research Center, VIB, Leuven, B-3000, Belgium; E-mail: peter.carmeliet@vib-kuleuven.be targeting of the FGFR by SSR has also shown therapeutic potential for the treatment of cancer and inflammatory diseases and represents a novel research tool to unravel the biology of RTK function.

1. Introduction

Receptor tyrosine kinases (RTK) belong to the enzyme-linked receptor superfamily and are proteins embedded in the cellular plasma membrane to serve the dual role of *recognition* of growth factor stimuli and *transduction* of these stimuli into cellular responses.[1] In the human genome, 58 RTKs act as high-affinity cell surface receptors, which are targeted by various inhibitors, including small molecule intracellular tyrosine kinase inhibitors (TKI), and extracellularly acting antibodies and peptides (Fig. 1). Currently, the majority of small molecule RTK inhibitors impair the tyrosine kinase (TK) activity as competitive antagonists by blocking binding of ATP to its substrate-binding site. Most kinase inhibitors that entered the clinic have however multiple additional (non-specific) off-target effects, thereby increasing the risk of adverse effects and sometimes requiring drug treatment holidays because of unacceptable toxicity.[2] Several alternative strategies are currently being evaluated, amongst which allosteric modulation of the Fibroblast Growth Factor receptors (FGFRs) show promising results.[3,4]

2. The FGF and FGFR Family

The Fibroblast Growth Factor receptor (FGFR) family is an important RTK family due to its widespread involvement in numerous physiological and pathological processes, and has therefore been targeted by multiple drug-discovery programs.[1,5] The FGF family comprises a large family of structurally-related polypeptide growth factor ligands, with members found in organisms from nematodes to humans (see Chapter 1 for details).[6] Most FGFs are broad-spectrum mitogens and stimulate various cellular functions including migration, proliferation, differentiation and stemness.[7] These activities are critical to a wide variety of physiological and pathological processes including angiogenesis, wound healing, tumorigenesis,

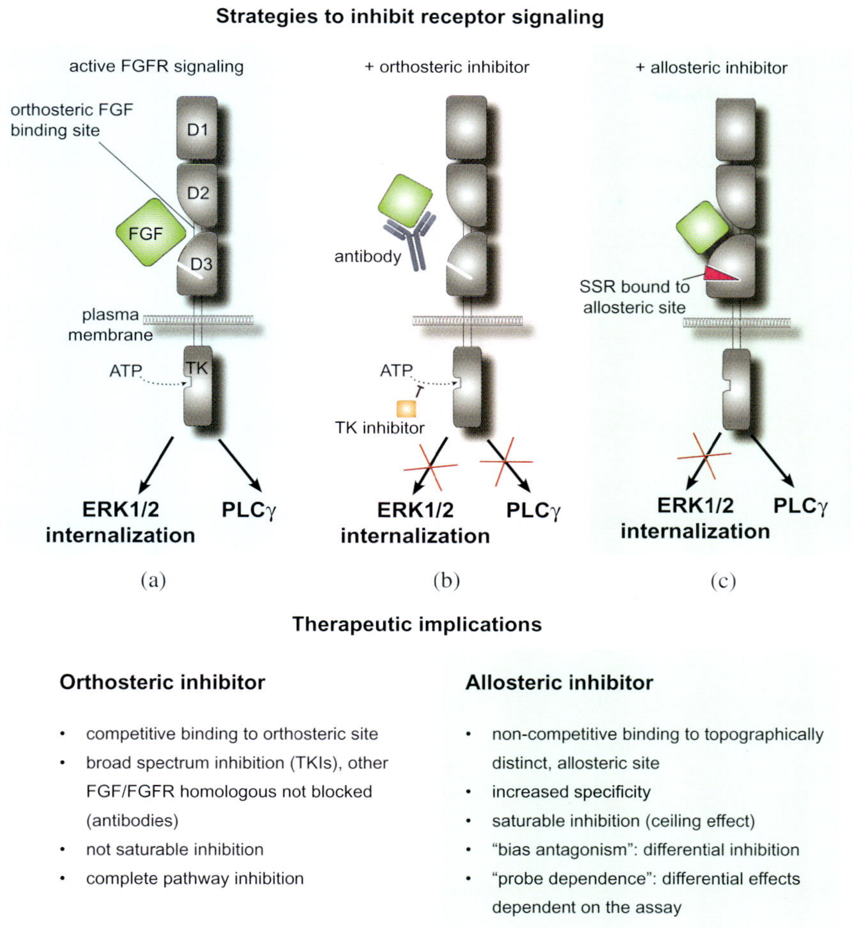

Fig. 1. Orthosteric vs. Allosteric modulation of the FGFR (a) Schematic representation of the FGFR (gray) which becomes activated by FGF-ligand binding (green) to the orthosteric binding site, thereby inducing downstream signaling. FGFRs can be targeted via orthosteric inhibitors (b) or allosteric modulators (c) Orthosteric inhibitors targeting the FGFR can be ligand traps (e.g. antibodies), which prevent FGF binding to the receptor or (ii) competitive antagonists, which compete for the same binding site on the receptor; both inhibitors eliminate the entire downstream response of the receptor. Allosteric modulators, such as SSR (pink) can modify the FGFR in various ways while still allowing the possibility of orthosteric FGF binding. Below the figure: comparison of the therapeutic features of orthosteric and allosteric inhibitors.[1]

development of the central nervous system (CNS), bone growth, metabolism and embryonic development.[5,8–12] In healthy adults, FGFs are involved in tissue repair, wound healing, and neuronal stem cell proliferation and migration.[13]

In mammals, the FGFs bind to four high affinity tyrosine kinase receptors (FGFR1-4).[14] In general, FGFRs are structured according to a conserved pattern (Fig. 1) consisting of (i) an extracellular part containing immunoglobulin (Ig)-like domains,[15] (ii) a helical, single-pass transmembrane domain that connects the extra- and intracellular parts, and (iii) an intracellular or cytosolic part, which contains the insert-split kinase domain that executes the transfer of the phosphate group to the target protein. The latter is flanked by a juxta-membrane (nearby the plasma membrane) domain and a carboxyterminal domain, which both contain regulatory phosphorylation sites and thereby regulate kinase activity.[15,16] The function of the kinase domain has been well described, although its activation and fine-tuning by the interaction with the extracellular and transmembrane domains remains largely unknown.

Differential splicing of FGFR transcripts expands the repertoire of these receptors. The different isoforms can contain an extracellular domain composed of either two or three Ig-like domains,[17] and alternative splicing in the third Ig-like domain can profoundly alter ligand-binding specificity.[7] The extracellular Ig domains are denoted D1, D2 and D3 (Fig. 1). Diversity in FGF signaling is achieved by the presence and spatial distribution of different FGFs, different FGFRs and their splice variants.

Each FGFR interacts with a subset of FGFs. When a ligand binds to its receptor, it changes the receptor's conformation to activate its biochemical signaling pathways. In other receptor classes, this change in conformation was thought to be actively promoted by the ligand (i.e., 'conformational induction'), but it is now known that most receptors are present as an ensemble of pre-existing states, and that the ligand preferentially stabilizes a state for which it has a higher affinity (i.e., 'conformational selection').[1] Whether this holds true for the FGFR still needs to be investigated.

What is known, is that the FGF ligand binds to an evolutionarily conserved 'orthosteric' ligand binding site located between domains

D2 and D3 (Fig. 1)[18] for which the high-resolution structure of the ligand-bound receptor was determined more than 15 years ago.[19–21] Still, the FGFRs remain one the best structurally characterized RTKs as for most other RTKs only subdomains of the extracellular domain have been structurally resolved. More than 30 amino acid residues in FGFR are involved in establishing contact with the FGF ligand.[19–21] Some residues of D3, making contact with FGF, are located in the differentially spliced region, thereby regulating differential ligand specificity for each splice form.[22]

3. FGFR Downstream Signaling

Signaling through the FGFRs occurs via ligand binding, which results in the dimerization of two FGFRs, thereby trans-phosphorylating each other on the intracellular tyrosine kinase domain. The presence of phosphorylated tyrosine residues makes it possible to recruit adaptor and docking molecules, which further maintain downstream signaling through a number of pathways, such as the phospholipase gamma (PLCγ) or the mitogen activated protein kinase (MAPK) or AKT pathway.[23] Heparin or heparan sulfate proteoglycans (HSPGs) act as low affinity receptors for FGFs and FGFRs, which rather function as stabilizing agents to enhance receptor dimerization and ligand presentation, and function as biological sinks for FGFs in the extracellular matrix.[24–26] Various co-receptors of FGFRs have been identified that alter FGFR signaling,[27] either by assembling intracellular signaling proteins or by interacting through various extracellular contact points. How FGF-ligands and co-receptors change the 'active' FGFR structure and how this affects downstream signaling remain some of the major outstanding questions.

4. Orthosteric Activation vs. Allosteric Modulation of Receptor Signaling

RTKs couple to downstream signaling pathways in a cell context-dependent manner.[1] Although the majority of studies on receptor biology have focused on how ligands mediate receptor signaling through binding to their orthosteric site, receptor conformation and signaling can also be modified by ligands acting at topographically and spatially distinct 'allosteric sites' (Fig. 1). The ability of a ligand (be it a small molecule or large protein) to change the properties of a second ligand (small molecule, protein) via an interaction with a non-overlapping and spatially distant site is referred to as allostery.[1,28,29] Apparently, nature has adopted this feature as a mechanism to fine-tune and regulate receptor function. The phenomenon of allostery has been gaining increasing attention recently as it is becoming evident that it can be used to modulate receptor function in ways that cannot be achieved by ligands that bind to an orthosteric site, making it an exciting new avenue in the design of novel therapeutics.[1,29]

Indeed, while orthosteric ligands compete for the binding to an agonist-binding site, which initiates downstream receptor signaling, allosteric ligands, which are usually structurally different from orthosteric ligands, non-competitively bind to spatially distinct sites from the orthosteric site. As such, allosteric agents can "modulate" the effect of an orthosteric ligand on its receptor. For instance, allosteric 'affinity modulators' can conformationally alter the orthosteric binding site leading to a different affinity of the orthosteric ligand for the receptor. 'Efficacy modulators' on the other hand, induce a conformational change that is transmitted to regions of the receptor involved in transduction of intracellular responses. This usually alters the signaling ability of the orthosteric ligand independently of, or in addition to, possible effects on orthosteric ligand affinity.

5. Allosteric Modulation of FGFR by SSR128129E

Recently, a novel small-molecule inhibitor, SSR128129E (from hereon abbreviated as "SSR"),[3,4] was identified as acting as an allosteric modulator of FGFRs by binding to the extracellular domain. As many other first-in-its-class allosteric modulators of other receptor classes,[1] SSR was identified serendipitously using an assay that was initially not designed to discover allosteric molecules.[3] Indeed, the initial screen was designed to identify competitive inhibitors in a scintillation proximity assay, in which the binding of radioactively labeled FGF to an

FGFR-Fc recombinant fusion protein was monitored. Here, SSR was identified as a weak inhibitor (μM affinity range), although in subsequent cellular response assays, it turned out to have nanomolar potency; this was a first hint for an allosteric mechanism. Indeed, from studies on other receptor classes, it is known that allosteric modulators can display "probe dependence", a phenomenon whereby the magnitude and/or direction of an allosteric effect can change markedly depending on the assay conditions or the chemical nature of the ligands occupying the orthosteric and allosteric site.[1] This can offer an advantage if the goal is to affect specifically one particular ligand-driven response, but may also pose a major hurdle to identify allosteric modulators, since the allosteric activities of a compound may only be identified in particular but not in all screening assays. As SSR was also not able to compete with FGF for its binding to the FGFR on the cell surface (as opposed to an Fc-coupled fusion receptor),[3] we further pursued its possible allosteric mode-of-action.

Another characteristic feature of allosteric molecules involves a higher "selectivity" for a particular receptor class or subtype as compared to orthosteric/competitive inhibitors. In contrast to orthosteric binding sites, allosteric binding sites are usually less conserved, up to the level where they can become unique for a single receptor subtype. In other instances, the allosteric site may be conserved across several subtypes of a receptor family, but selectivity is still obtained because the allosteric effect (be it positive or negative) only occurs in the receptor subtype of interest while remaining 'neutral' for other receptor subtypes.[1,30] By performing binding assays of additional closely and more distantly related RTKs and their respective ligands, SSR was found to be incapable of inhibiting their respective binding.[3] Also, in several assays, screening the functional activity of these other RTKs, including phosphorylation and various cellular responses, SSR turned out to selectively inhibit FGF responses, while not affecting other RTKs.[3]

A third, initially puzzling observation, was the inability of SSR to inhibit FGFR phosphorylation completely upon stimulation with FGF. This finding suggested a phenomenon of "saturability". As SSR does not compete with the FGF ligand for binding to the FGF receptor, the effective dose of SSR reaches a 'saturable' (ceiling) level. Once every receptor molecule is ligated, there is no additional pharmacological effect even in the presence of increased concentrations of the allosteric modulator. This implies that allosteric modulators have a greater potential for on-target safety in overdose situations than orthosteric modulators. In contrast, to maintain on-target efficacy, the levels of molecules that competitively inhibit orthosteric ligand binding should be increased if orthosteric ligand levels increase, but such high doses of orthosteric inhibitors may then also induce off-target toxicity.[28]

A fourth feature of allosteric modulators is their ability to impart "biased antagonism". This phenomenon beholds that in the presence of an allosteric modulating agent, downstream signaling can be altered, whereby one particular pathway can be stimulated or inhibited, while another remains unaffected (or even shows an opposite effect). When analyzing the effect of SSR on the various FGFR downstream signaling pathways, we observed that the FGFR substrate 2 (FRS2) and Extracellular signal-Regulated Kinase (ERK) signaling cascade was efficiently inhibited upon FGF stimulation, while the phospholipase C (PLC)-γ pathway remained unaffected (Figs. 1 and 2).[3,4] Compounds displaying such biased antagonism may offer a therapeutic advantage, as they can block a key pathway promoting the progression of a particular disease, while not affecting other pathways that are necessary for homeostatic maintenance and would cause adverse effects when blocked.

Overall, the combination of probe dependence of SSR, the high selectivity towards the FGFR family, the non-competitive binding to the FGFR and the biased antagonism on FGFR downstream signaling, indicates that SSR functions as an allosteric efficacy modulator of the FGFRs, and suggests that these pharmacological features could be used in future screening efforts to find other and different FGFR modulators.

6. Molecular Mechanism of SSR128129E

The next crucial step was to determine how SSR mediates its allosteric modulation of FGFR on the molecular level. To unravel the

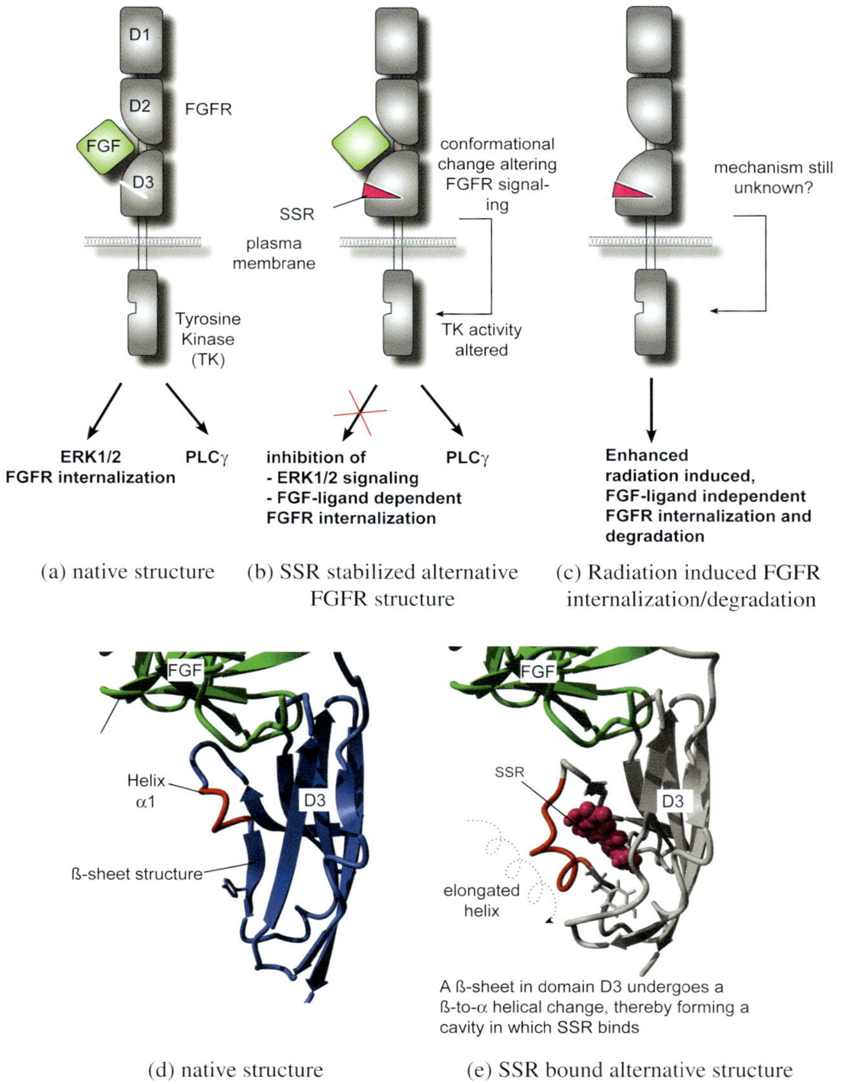

Fig. 2. Molecular mechanism of SSR, a small molecule allosteric FGFR inhibitor (a) Scheme illustrating the native structure of the fibroblast growth factor receptor (FGFR, gray), containing three extracelluar domains (D1, D2 and D3) and the intra-cellular tyrosine kinase domain. The orthosteric binding site for the FGF ligand (green) is located between domain 2 (D2) and domain 3 (D3). Ligand binding activates downstream signaling through ERK1/2 and PLCγ, and leads to FGF-ligand

Fig. 2. (***Continued***) dependent receptor internalization. (b) Scheme illustrating the FGFR bound to the small molecule allosteric inhibitor SSR129128E (SSR; red), binding to a distant site from the orthosteric site in domain D3. This stabilizes a structural change in domain D3, which selectively inhibits signaling through ERK1/2 and internalization, without affecting PLCγ signaling. (c) Upon the combined treatment of SSR and irradiation, the FGFR was internalized more efficiently and degraded, leading to a enhanced sensitivity of the tumor cells to irradiation. (d) High magnification 3 dimensional structure of the orthosteric ligand binding site. The α helix and adjacent ß-helix (prone to extending into an α helix) in domain D3 are highlighted in red and blue, respectively. (e) The FGFR can undergo a structural change involving a ß-to-α helical change (i.e. the α-helix extends by 2 twists, and is therefore referred to as the 'H2-conformation') of the labile ß-sheet in domain D3, thereby forming a cavity in which SSR can bind with high affinity. As such, SSR stabilizes a different receptor conformation, which affects downstream signaling.[1,4]

molecular mechanism, we used a combination of a broad array of state-of-the-art biophysical, computational and biological methods, including crystallography, two-dimensional nuclear magnetic resonance, Fourier transform infrared spectroscopy, molecular dynamics simulations, free energy calculations, structure-activity relationship analysis and FGFR mutagenesis.[5] Overall, the allosteric mechanism of inhibition relies on the binding of SSR to a highly FGFR-specific pocket in extracellular domain D3, adjacent to the cell membrane, which is non-overlapping with the orthosteric binding site of the FGF ligand (Fig. 2).[5] This binding pocket becomes available through the normal conformational fluctuation of a stretch with moderate ß-sheet features that can be shortened by the extension of an adjacent α-helix.[5] Via a high affinity binding of SSR in this newly formed cavity, the FGFR becomes stabilized into an alternative conformation. This structural change seems sufficient to alter the signaling properties of the FGFR. While the FRS2/ERK1/2 pathway is inhibited by SSR, the PLC-γ pathway remains unaffected (Figs. 1 and 2).[3]

7. Therapeutic Implications of Allosteric FGFR Modulation

7.1. *Anti-tumoral activities of SSR*

FGFs and FGFRs play a major role during cancer development and progression.[5] Not only do they contribute to tumor growth, metastasis, tumor angiogenesis and inflammation, they also contribute to tumor resistance mechanisms to radiotherapy[31]. Increasing evidence illustrates that allosteric modulation of the FGFR is an effective novel strategy for various pathological conditions.

A first anti-tumoral mechanism of SSR is aimed at the inhibition of FGFR signaling in tumor and stromal cells that require FGFR and MAPK signaling. This was analyzed in various pre-clinical syngeneic and orthotopic tumor models, and human xenograft tumor models, which are to some extent all dependent on FGFR signalling. Oral delivery of SSR inhibited or delayed the growth of orthotopic Panc02 pancreatic tumors, subcutaneous Lewis lung carcinoma (LLC), 4T1 breast cancer cells, CT26 colon tumor cells in syngenic mice, and the growth of a multidrug resistant MCF7/ADR breast cancer cell line in a xenograft model.[3] In addition, SSR reduced tumor invasiveness, inhibited metastasis to draining lymph nodes and reduced the number of metastatic nodules in the lung.[3]

The FGF/FGFR axis is also an important player in tumor angiogenesis (i.e. the growth of tumor associated blood vessels). Anti-angiogenic therapy was suggested to be a promising treatment paradigm, although a substantial fraction of cancer patients does not or only minimally responds to anti-angiogenic inhibitor therapy, such as those targeting the vascular endothelial growth factor (VEGF) and its receptors.[32] This refractoriness is in part due to the upregulation of additional angiogenic growth factors such as members of the FGF family. Combinatorial treatment of SSR with anti-VEGFR was more effective at inhibiting tumor growth as compared to monotherapy with each agent alone,[3] showing that combination therapies are indeed a valuable strategy.

In addition, subcutaneous injection of Lewis Lung carcinoma cells into mice increased the amount of circulating endothelial progenitor cells (EPCs) from peripheral blood, an effect that is associated with enhanced tumor growth and an increase of the intra-tumoral vascularization.[33] Treatment with SSR decreased the amount of circulating EPCs from the peripheral blood demonstrating the involvement of the FGF/FGFR axis in EPC recruitment during angiogenesis-dependent tumor growth.[33]

Recent evidence indicates that FGFR expression in glioblastoma patients serves as a prognostic factor of survival and time to progression after radio-chemotherapy and was suggested as a key mechanism of resistance.[31] Subsequent research revealed that allosteric modulation of FGFR by SSR can change the cellular localization of FGFR upon tumor cell irradiation. Indeed, treatment of radio-resistant glioblastoma cells with SSR, radio-sensitized cancer cells by inducing mitotic cell death.[34] This effect was achieved by decreasing the cell membrane availability of FGFR by increasing receptor endocytosis and ubiquitin dependent degradation, and modulation of FGFR-dependent hypoxia signaling pathways.[34] Treating glioblastoma xenograft mouse models with the combination of SSR and radiotherapy significantly improved the outcome in preclinical models.[34] This strongly suggests that allosteric targeting of the FGFR by SSR represents an interesting strategy to improve the efficiency of radiotherapy.

This last example highlights the complexity of allosteric modulators and how the outcome depends on the cellular and assay context. Indeed, in case of radiotherapy, SSR increased FGF-ligand independent internalization of cell surface FGFR resulting in an enhanced sensitivity of the tumor cells to radiotherapy, while in FGF-ligand dependent cellular models, SSR inhibited FGF2-induced endocytosis of FGFR in early endosomal antigen1 positive (EEA1+) vesicles.[4] While in both cases, anti-tumoral effects of SSR could be observed, this highlights that different tumors can exhibit a broad variety of responses upon allosteric FGFR modulation. A deeper insight into receptor structure-function and signaling will be required to fully grasp the complexity of these observations.

7.2. *Anti-inflammatory and anti-atherosclerotic activities of SSR*

Inflammatory diseases, such a rheumatoid arthritis, are also known to be FGFR-driven processes.[35] Oral administration of SSR reduced the severity of the clinical symptoms (i.e. number of limbs affected by redness, swelling and deformity) in a mouse model of acute rheumatoid arthritis and slowed down the disease progression.[3] In this model, SSR reduced FGF-related angiogenesis in the inflamed joints and attenuated synovial hyperplasia, inflammatory cell infiltration and cartilage breakdown.[3]

Also in an atherosclerosis-prone, apolipoprotein E (apoE)-deficient mouse model, treatment with SSR decreased the proliferation of blood vessels and reduced the inflammatory cell influx[36] — both process that are crucial in the formation of atherosclerotic plaques.[37] SSR treatment also resulted in a reduction of the lesion sizes, without any change in serum lipids. This indicates that FGFR signaling has an important role in the development of vascular diseases such as atherosclerosis and shows that allosteric modulation of the FGFR offers a new therapeutic approach.

8. Safety Profile

Finally, in all these pre-clinical mouse models, SSR, and therefore also the allosteric modulation of FGFRs, was generally well tolerated. This is in line with the overall lower toxicity of allosteric inhibitors (cf. infra). At therapeutic doses in adult mice, SSR did not affect blood vessel stability, arterial blood pressure or thrombotic events.[3] Hematological parameters were mostly normal nor did treatment with SSR affect the body weight or alter plasma cholesterol and triglyceride levels.[3,38] Identifying the disease related downstream signalling pathways and developing specific inhibitors seems therefore a safe therapeutic strategy.

9. Conclusions

The variety of therapeutic effects of a novel allosteric modulating compound targeting FGFRs highlights the enormous opportunities that are linked to allosteric RTK modulation. It is also becoming clear that allosteric modulators, being it small chemical compounds, peptides or antibodies, are very attractive therapeutics and several of those have and are being identified to target other members of the RTK family.[1] However, based on the biological observations made in the presence of allosteric modulators, it is also becoming clear that our knowledge of cellular receptor function is still very limited in the context of intact cells, let be entire organs and organisms. Nevertheless, some other allosteric modulators of the FGFRs are on the way, including another small molecule inhibitor targeting FGFR2,[39] showing that this strategy is gaining traction in drug discovery programs.

Acknowledgements

F.D.S. is supported by the Research Foundation Flanders (FWO). This work is supported, by grant #G.0789.11 and #G.00764.10 from the FWO, Belgium; the Belgian Science Policy (IAP #P7/03); Leducq Network of Excellence; and long-term structural Methusalem funding by the Flemish Government to P.C.

References

1. De Smet, F., Christopoulos, A. and Carmeliet, P. Allosteric targeting of receptor tyrosine kinases. *Nat Biotechnol* **32**, 1113–1120 (2014).
2. Shah, D.R., Shah, R.R., and Morganroth, J. Tyrosine kinase inhibitors: Their on-target toxicities as potential indicators of efficacy. *Drug Saf* **36**, 413–426 (2013).
3. Bono, F., *et al.* Inhibition of tumor angiogenesis and growth by a small-molecule multi-FGF receptor blocker with allosteric properties. *Cancer Cell* **23**, 477–488 (2013).
4. Herbert, C., *et al.* Molecular mechanism of SSR128129E, an extracellularly acting, small-molecule, allosteric inhibitor of FGF receptor signaling. *Cancer Cell* **23**, 489–501 (2013).
5. Herbert, C., Lassalle, G., Alcouffe, C., and Bono, F. Approaches targeting the FGF-FGFR system: A review of the recent patent literature and associated advanced therapeutic agents. *Pharm Pat Anal* **3**, 585–612 (2014).
6. Ornitz, D.M., and Itoh, N. Fibroblast growth factors. *Genome Biol* **2**, ReviewS3005 (2001).
7. Eswarakumar, V.P., Lax, I., and Schlessinger, J. Cellular signaling by fibroblast growth factor receptors. *Cytokine Growth Factor Rev* **16**, 139–149 (2005).

8. Powers, C.J., McLeskey, S.W., and Wellstein, A. Fibroblast growth factors, their receptors and signaling. *Endocr Relat Cancer* **7**, 165–197 (2000).
9. Klint, P., and Claesson-Welsh, L. Signal transduction by fibroblast growth factor receptors. *Front Biosci* **4**, D165–D177 (1999).
10. Goldfarb, M. Functions of fibroblast growth factors in vertebrate development. *Cytokine Growth Factor Rev* **7**, 311–325 (1996).
11. Presta, M., *et al.* Fibroblast growth factor/fibroblast growth factor receptor system in angiogenesis. *Cytokine Growth Factor Rev* **16**, 159–178 (2005).
12. Reuss, B., and von Bohlen und Halbach, O. Fibroblast growth factors and their receptors in the central nervous system. *Cell Tissue Res* **313**, 139–157 (2003).
13. Coutu, D.L., and Galipeau, J. Roles of FGF signaling in stem cell self-renewal, senescence and aging. *Aging (Albany NY)* **3**, 920–933 (2011).
14. Gerwins, P., Skoldenberg, E., and Claesson-Welsh, L. Function of fibroblast growth factors and vascular endothelial growth factors and their receptors in angiogenesis. *Crit Rev Oncol Hematol* **34**, 185–194 (2000).
15. Hubbard, S.R. Autoinhibitory mechanisms in receptor tyrosine kinases. *Front Biosci* **7**, d330–d340 (2002).
16. Zhang, J., Yang, P.L., and Gray, N.S. Targeting cancer with small molecule kinase inhibitors. *Nat Rev Cancer* **9**, 28–39 (2009).
17. Wilson, B.D., *et al.* Netrins promote developmental and therapeutic angiogenesis. *Science* **313**, 640–644 (2006).
18. Neubig, R.R., Spedding, M., Kenakin, T., and Christopoulos, A. International Union of Pharmacology Committee on Receptor Nomenclature and Drug Classification. XXXVIII. Update on terms and symbols in quantitative pharmacology. *Pharmacol Rev* **55**, 597–606 (2003).
19. Pellegrini, L., Burke, D.F., von Delft, F., Mulloy, B., *et al.* Crystal structure of fibroblast growth factor receptor ectodomain bound to ligand and heparin. *Nature* **407**, 1029–1034 (2000).
20. Plotnikov, A.N., Schlessinger, J., Hubbard, S.R., and Mohammadi, M. Structural basis for FGF receptor dimerization and activation. *Cell* **98**, 641–650 (1999).
21. Stauber, D.J., DiGabriele, A.D., and Hendrickson, W.A. Structural interactions of fibroblast growth factor receptor with its ligands. *Proc Natl Acad Sci USA* **97**, 49–54 (2000).
22. Mohammadi, M., Olsen, S.K., and Ibrahimi, O.A. Structural basis for fibroblast growth factor receptor activation. *Cytokine Growth Factor Rev* **16**, 107–137 (2005).

23. Beenken, A., and Mohammadi, M. The FGF family: Biology, pathophysiology and therapy. *Nat Rev Drug Discov* **8**, 235–253 (2009).
24. Schlessinger, J., *et al.* Crystal structure of a ternary FGF-FGFR-heparin complex reveals a dual role for heparin in FGFR binding and dimerization. *Mol Cell* **6**, 743–750 (2000).
25. Lin, X., Buff, E.M., Perrimon, N., and Michelson, A.M. Heparan sulfate proteoglycans are essential for FGF receptor signaling during Drosophila embryonic development. *Development* **126**, 3715–3723 (1999).
26. Ornitz, D.M., FGFs, Heparan sulfate and FGFRs: Complex interactions essential for development. *Bioessays* **22**, 108–112 (2000).
27. Murakami, M., Elfenbein, A., and Simons, M. Non-canonical fibroblast growth factor signalling in angiogenesis. *Cardiovasc Res* **78**, 223–231 (2008).
28. Christopoulos, A. Allosteric binding sites on cell-surface receptors: Novel targets for drug discovery. *Nat Rev Drug Discov* **1**, 198–210 (2002).
29. Wootten, D., Christopoulos, A., and Sexton, P.M. Emerging paradigms in GPCR allostery: Implications for drug discovery. *Nat Rev Drug Discov* **12**, 630–644 (2013).
30. Birdsall, N.J., and Lazareno, S. Allosterism at muscarinic receptors: Ligands and mechanisms. *Mini Rev Med Chem* **5**, 523–543 (2005).
31. Ducassou, A., *et al.* alphavbeta3 Integrin and Fibroblast growth factor receptor 1 (FGFR1), Prognostic factors in a phase I-II clinical trial associating continuous administration of Tipifarnib with radiotherapy for patients with newly diagnosed glioblastoma. *Eur J Cancer* **49**, 2161–2169 (2013).
32. Ebos, J.M., and Kerbel, R.S. Antiangiogenic therapy: Impact on invasion, disease progression, and metastasis. *Nature Reviews. Clinical Oncology* **8**, 210–221 (2011).
33. Fons, P., *et al.* Tumor vasculature is regulated by FGF/FGFR signaling-mediated angiogenesis and bone marrow-derived cell recruitment: This mechanism is inhibited by SSR128129E, the first allosteric antagonist of FGFRs. *J Cell Physiol* **230**, 43–51 (2015).
34. Ader, I., *et al.* Preclinical evidence that SSR128129E — a novel small-molecule multi-fibroblast growth factor receptor blocker — radiosensitises human glioblastoma. *Eur J Cancer* **50**, 2351–2359 (2014).
35. Malemud, C.J., Growth hormone, VEGF and FGF: Involvement in rheumatoid arthritis. *Clin Chim Acta* **375**, 10–19 (2007).

36. Dol-Gleizes, F., *et al.* A new synthetic FGF receptor antagonist inhibits arteriosclerosis in a mouse vein graft model and atherosclerosis in apolipoprotein E-deficient mice. *PLoS One* **8**, e80027 (2013).
37. Yla-Herttuala, S., *et al.* Stabilization of atherosclerotic plaques: An update. *Eur Heart J* **34**, 3251–3258 (2013).
38. Scroyen, I., Vranckx, C., and Lijnen, H.R. FGF receptor antagonism does not affect adipose tissue development in nutritionally induced obesity. *Adipocyte* **3**, 46–49 (2014).
39. Tsimafeyeu, I., *et al.* Allosteric FGFR2 inhibitor RPT835 impacts on tumor growth and neovascularization. *Research,* **74**, 1737–1737 (2014).

Chapter 8

Fibroblast Growth Factors in Epithelial Homeostasis and Repair

Michael Meyer*, Luigi Maddaluno* and Sabine Werner†

Abstract

Maintenance of epithelial integrity is essential for tissue and whole body homeostasis and for prevention of various types of diseases, including acute and chronic inflammatory diseases, fibrosis and cancer. Therefore, injuries to epithelial tissues have to be rapidly and efficiently repaired. FGFs are key regulators of epithelial development, and recent data demonstrate that they also control epithelial homeostasis and the response to various types of injury. Here we summarize the current knowledge on the role of endogenous FGFs and their receptors in homeostasis and repair of selected epithelial tissues, for which the function of FGFs has been particularly well studied.

1. Introduction

Fibroblast growth factors (FGFs) comprise a family of 22 proteins that regulate survival, migration, proliferation, and/or differentiation of

Institute of Molecular Health Sciences, Department of Biology, ETH Zurich, Otto-Stern-Weg 7, 8093 Zurich, Switzerland.

*Equal contribution.

†Corresponding author: sabine.werner@biol.ethz.ch

many different cell types (see Chapter 1) (reviewed by[1,2]). They are master regulators of organogenesis and tissue homeostasis, and mutations in *FGF* or FGF receptor (*FGFR*) genes cause developmental/genetic diseases involving various organs, in particular the bone, but also the skin (reviewed by[1,2]). In addition, abnormal expression of FGFs or their receptors has been demonstrated in different types of benign tumors and malignant cancers, and FGFR inhibitors are therefore in clinical trials for the treatment of some types of malignancies (reviewed by[3,4]). FGFs exert their functions by activating four transmembrane tyrosine kinase receptors, designated FGFR1-4 (reviewed by[2]). Further complexity is achieved by alternative splicing, e.g. in the part of the RNA that encodes the third immunoglobulin-like domain of FGFR1-3. These splicing events generate IIIb and IIIc variants with different ligand binding specificities. The IIIb splice variants are almost exclusively expressed in epithelial cells, whereas mesenchymal and immune cells express predominantly the IIIc variants (reviewed by[2,5]). The IIIb splice variant of FGFR2 (FGFR2b) is a high affinity receptor for FGF1, –3, –7, –10, and –22, whereas the IIIc splice variant of this receptor (FGFR2c) is activated by another set of FGFs. The ligand binding specificity of FGFR1b is similar to that of FGFR2b, although FGFR1b does not bind FGF7. By contrast, the IIIb variant of FGFR3 binds mainly FGF1, –9, –16, and –20.[6] These ligand binding specificities indicate overlapping functions of FGFR1b and FGFR2b, but distinct functions of FGFR3b. They also show that epithelial cells respond to a different set of FGFs compared to stromal cells. Here we summarize data demonstrating the importance of epithelial FGFR signaling for tissue homeostasis and repair in adult mammals with a focus on the epithelial tissues, where FGF function has been most intensively studied. We further focus our article on the function of endogenous FGFs and FGF receptors, while the various therapeutic effects of recombinant FGFs are reported elsewhere (reviewed by[5,7,8]).

2. Role of FGF Signaling in Keratinocytes for Skin Homeostasis and Repair

Keratinocytes of the skin express FGFR1-3 and thereof predominantly or even exclusively the IIIb variants.[9,10] The role of endogenous

FGFR3 in keratinocytes for skin homeostasis and repair is as yet unclear, since FGFR3 knockout mice lack obvious skin abnormalities (S. Werner and D.M. Ornitz, unpublished). However, reduced levels of FGFR3 and FGFR2 were found in the epidermis of mice, which carry a mutation in the gene encoding the p63 transcription factor, which causes ankyloblepharon-ectodermal defects-cleft lip/palate (AEC) syndrome that is characterized by a severe skin defect due to a defective epidermal stem cell compartment. The proliferative capacity of the keratinocytes from these mice was rescued by overexpression of FGFR2b.[11]

Activating mutations in the *FGFR3* gene were found in patients with *acanthosis nigricans*, a skin disease associated with hyperkeratosis and hyperpigmentation,[12] and somatic activating mutations are present in epidermal nevi[13] and in seborrheic keratosis.[14] Interestingly, activation of FGFR3 promotes keratinocyte differentiation, a finding that correlates with the predominant expression of this receptor in benign *versus* malignant skin tumors.[15]

Deletion of the IIIb exon of FGFR2 in all cells caused perinatal lethality of the knockout mice, and they showed epidermal hypoplasia and hair follicle abnormalities.[16] Similar skin abnormalities were observed in transgenic mice expressing a dominant-negative FGFR2b mutant in keratinocytes, although these mice are viable since the mutant receptor is mainly expressed in keratinocytes.[17] Mice lacking one of the FGFR2b ligands, FGF10, also had a thin epidermis, whereas the hair follicles were not affected.[16,18] FGF7 (keratinocyte growth factor; KGF) knockout mice were characterized by rough fur, resulting from a perturbation in the orientation of the hairs,[19] but they did not exhibit other skin abnormalities. Finally, mice lacking FGFR2b specifically in keratinocytes developed cutaneous inflammation upon aging and revealed a higher susceptibility to chemically-induced skin carcinogenesis.[20] However, the defects were rather mild in young mice under unchallengend conditions, suggesting functional redundancy among different FGF receptors. To test this possibility, mice lacking FGFR1 and FGFR2 in keratinocytes were generated by mating mice with floxed *FGFR1* and *FGFR2* alleles with transgenic mice expressing Cre recombinase under the control of the keratin 5 (K5) promoter. The progeny of this breeding included mice lacking FGFR1, FGFR2, or both receptors in keratinocytes (designated K5-R1, K5-R2 and K5-R1/R2 mice). Whereas no phenotype was observed in K5-R1 mice, K5-R2 mice developed the same phenotype as mice lacking only the FGFR2b variant in keratinocytes,[10] including loss of sebaceous glands, hair abnormalities and defects in the external genitalia. A severe phenotype was observed in K5-R1/R2 mice: All females and more than 60% of the males were infertile and they had a reduced body size and weight. Nevertheless, the animals have a normal life expectancy. They progressively lost their hair and were bald by the age of 2–4 months. This phenotype most likely results from a failure to regenerate hair follicles, a finding that is consistent with the proposed role of FGF7 and FGF10 in hair follicle regeneration. Thus, expression of FGF7 and FGF10 increases in the dermal papilla of the hair follicle during the transition from the early to the late telogen, and these FGFs are highly expressed at this site during the anagen phase of the hair cycle.[21,22] Functional *in vitro* studies revealed that FGF7 and FGF10 stimulate proliferation of hair germ cells adjacent to the dermal papilla, resulting in hair cycle activation.[22] Therefore, the progressive hair loss in K5-R1/R2 mice is consistent with all these observations.

A role of FGF18 in hair follicle cycling has also been demonstrated. This type of FGF is expressed in a stem cell niche of the hair follicle during the telogen phase. Upon loss of FGF18 in keratinocytes, the telogen phase was strongly shortened, resulting in rapid succession of hair cycles.[23] Consistent with this finding it was demonstrated that FOXP1, a transcription factor that is crucial for maintaining the quiescence of hair follicle stem cells, exerts this effect by controlling FGF18 expression.[24] Thus, FGF18, which signals via FGFR3 or FGFR4,[6] is important for telogen maintenance, whereas FGF7 and FGF10 promote anagen through signaling via FGFR1b and FGFR2b.

In addition to the hair loss, K5-R1/R2 mice developed a progressive inflammatory skin disease with hyperproliferation of keratinocytes and epidermal thickening (acanthosis) as a result of keratinocyte hyperproliferation.[10] The same phenotype was seen in mice lacking

FGFR1b in all cells and FGFR2 in keratinocytes, demonstrating that FGFR1b is the responsible receptor splice variant.[10] Since FGFR1/FGFR2-deficient keratinocytes proliferated normally in culture, the hyperproliferation seen *in vivo* is not cell autonomous, but rather results from the progressive skin inflammation. This inflammatory response is caused by a defect in the epidermal barrier as shown by increased transepidermal water loss. This resulted from a reduction in tight junction protein expression in the knockout animals and consequent tight junction abnormalities that preceded the onset of inflammation. The effect on tight junction proteins is most likely the primary defect in these mice, since reduced expression of tight junction proteins was also seen in cultured keratinocytes. Furthermore, FGF7 induced the expression of the genes encoding these tight junction proteins in keratinocytes from wild-type mice. Consistent with the proposed defect in tight junctional permeability, the transepithelial electrical resistance of cultured keratinocytes from K5-R1/R2 mice was strongly reduced.[10] The important effect of FGFs on epidermal barrier function was also demonstrated by the finding that a small peptide derived from the receptor-binding domain of FGF2 enhanced the transepithelial electrical resistance of γ-irradiated human keratinocytes grown on transwells through upregulation of tight junction and adherens junction components.[25]

The first signs of inflammation in K5-R1/R2 mice were the accumulation of dermal mast cells and of epidermal γδ T cells and their activation,[26,27] and these alterations occurred prior to the development of the epidermal abnormalities. Therefore, the importance of these immune cells for the phenotype was determined by generation of triple mutant mice lacking FGFR1 and FGFR2 in keratinocytes as well as mast cells or γδ T cells. Surprisingly, both types of triple mutant mice had similar abnormalities as K5-R1/R2 mice,[26,27] demonstrating that neither mast cells nor γδ T cells play a causative role in the hyperproliferative/inflammatory skin phenotype.

The skin abnormalities that develop in K5-R1/R2 mice are strongly reminiscent to those that develop in patients suffering from the inflammatory skin disease Atopic Dermatitis (AD).[26] This raises the intriguing question of a potential role of abnormal FGFR

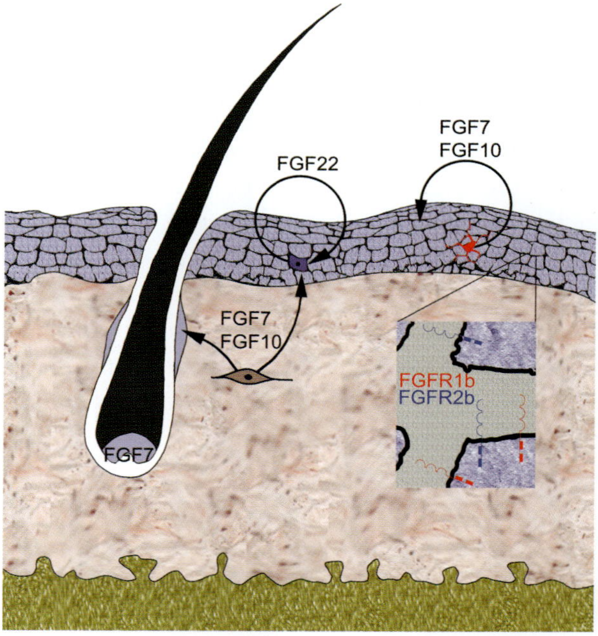

Fig. 1. The FGF7 family and its receptors in the skin. FGF7 and FGF10 are produced by fibroblasts of the dermis and the dermal papilla and they act in a paracrine manner on keratinocytes of the epidermis and the hair follicles, which express FGFR1b and FGFR2b. FGF22 is expressed in keratinocytes and acts in an autocrine manner on the same cell type via FGFR1b and FGFR2b. FGFR3b, which is also expressed in keratinocytes, but not shown in this figure, responds to another set of ligands.

signaling in the pathogenesis of this common human disease. Consistent with such a function, expression of FGFR1 and FGFR2 was strongly down-regulated in the skin of a subset of AD patients,[28] although it remains to be determined if this down-regulation occurs in keratinocytes or in other cell types of the skin. While Atopic Dermatitis may be associated with a deficiency in FGFR signaling, enhanced activation of FGFR1b and FGFR2b could be at least in part responsible for the hyperproliferation of keratinocytes that occurs in Psoriasis, since expression of its ligands FGF7 and FGF10 was upregulated in the upper dermis of patients suffering from this inflammatory skin disease.[29,30] Consistent with a role of FGFR2b in Psoriasis, overexpression of FGF7 in keratinocytes of transgenic mice induced a hyperproliferative skin phenotype.[31]

In addition to its role in epidermal homeostasis and hair follicle regeneration, FGFs are potent regulators of wound reepithelialization. This was first suggested by the strong upregulation of FGF7/KGF expression in fibroblasts of the granulation tissue and the dermis adjacent to skin wounds in response to cutaneous injury in mice and humans.[32,33] The increase is most likely a result of high levels of serum growth factors and pro-inflammatory cytokines present at the wound site, since these factors induce expression of FGF7 in cultured keratinocytes.[34,35] In addition, FGF7 is expressed in epidermal γδ T cells and upregulated in these cells upon wounding.[36,37] Consistent with an important role of FGF7 in wound repair, the upregulation of FGF7 that occurs in normal mice was much less pronounced in mice with delayed wound healing, including genetically diabetic mice and glucocorticoid-trated mice.[38,39] Surprisingly, however, FGF7-deficient mice did not exhibit obvious wound healing abnormalities in an incisional wound healing model,[19] although the delayed healing of excisional wounds in diabetic mice was further aggravated by loss of FGF7.[40] The normal healing in healthy mice suggested functional redundancy among different FGF family members. Indeed, expression studies revealed that FGF10 and FGF22 are also expressed in normal skin and upregulated after skin wounding.[41–44] Like FGF7, FGF10 is expressed by fibroblasts and γδ T cells,[37,41] whereas FGF22 is expressed by keratinocytes and most likely acts in an autocrine manner.[43] Redundancy among FGF7 and FGF10 in healing wounds is supported by the finding that mice lacking γδ T cells show a defect in wound reepithelialization, most likely due to the loss of epidermal γδ T cell-derived FGF7 and FGF10.[37] The consequences of the concomitant loss of FGF7 and FGF10 for wound healing could not be studied, since FGF10 knockout mice die immediately after birth.[45,46] FGF22 knockout mice did not exhibit a defect in the healing of full-thickness excisional wounds.[47] It will, however, be interesting to determine the consequences of a combined loss of

FGF7 and FGF22 on the wound healing process. An important role of FGFR signaling in keratinocytes for wound reepithelialization as well as redundancy among different FGFs and FGFRs was further supported by the phenotype of mice expressing a dominant-negative FGFR mutant in keratinocytes. These animals showed a severe delay in wound reepithelialization.[17] However, the responsible receptor(s) could not be identified with this approach, since the dominant-negative mutant blocks the action of all FGFRs in response to common ligands. To identify the responsible receptor(s), K5-R1, K5-R2 and K5-R1/R2 mice, which lack either FGFR1 or FGFR2 or both receptors in keratinocytes, were subjected to full-thickness excisional wounding. While the single mutant mice showed normal wound repair, wound closure was severely delayed in the double knockout mice due to impaired wound contraction and reepithelialization. The latter resulted from impaired keratinocyte migration at the wound edge. This deficiency was also seen *in vitro*, and FGFR1/2-deficient keratinocytes had a reduced migration velocity and impaired directional persistence as a result of inefficient formation and turnover of focal adhesions.[48,49] Mechanistically, this was caused by a significant reduction in the expression of major focal adhesion components in the absence of FGFR1 and FGFR2, resulting in a general migratory deficiency. In contrast to keratinocyte migration, the proliferative capacity of the wound keratinocytes was not affected, but rather enhanced as already seen in non-wounded skin of these mice.[48] Since a defect in wound reepithelialization that is characterized by impaired keratinocyte migration, but enhanced proliferation, is also a characteristic feature of chronic, non-healing skin ulcers in humans,[50] a potential role of defective FGFR signaling in the pathogenesis of such ulcers should be investigated in the future.

While these studies show an important role of FGFR1b and FGFR2b ligands for wound reepithelialization, FGF9, which does not activate these receptors, was identified as a promoter of hair follicle neogenesis after wounding. Upon excisional skin wounding of mice, FGF9 was initially produced by dermal γδ T cells and subsequently by wound fibroblasts, and its overexpression promoted hair follicle neogenesis.[51] It remains to be determined if this effect is directly mediated via keratinocytes, e.g. via FGFR3b, or indirectly via other cell types in the wound.

While activation of FGFR1b and FGFR2b ligands drives wound reepithelialization, other members of the FGF family are crucial regulators of wound angiogenesis. Thus, application of FGF2 neutralizing antibodies to skin wounds in wild-type rats and wound healing studies with FGF2 knockout mice revealed a role of FGF2 in the early phase of wound angiogenesis and in granulation tissue formation in general.[52,53] The phenotype was not aggravated by concomitant loss of FGF1, demonstrating that FGF1 does not compensate for the lack of FGF2 in healing skin wounds.[54] Consistent with an important role of FGF2 in wound angiogenesis, mice lacking FGFR1 and FGFR2 in endothelial cells and hematopoietic cells showed impaired neovascularization after skin wounding, which was associated with delayed wound repair.[55] Therefore, FGFR signaling in cells of the granulation tissue seems to affect epithelial repair in an indirect manner.

Whereas loss of FGF signaling is detrimental for wound healing, enhancement of FGF activity strongly promoted the wound repair process as demonstrated by local application of wild-type or modified FGF7 or other FGFs to different types of skin wounds or through enhancement of the levels of these proteins in wounds via gene therapy approaches (reviewed by[5,8]). In addition, FGF signaling can be promoted in healing wounds by FGF binding protein 1 (FGF-BP1). This secreted protein binds FGF family members and enhances their activities by facilitating their release from the extracellular matrix, allowing their access to the signaling receptors. When FGF-BP1 was inducibly overexpressed in transgenic mice under control of a ubiquitously active promoter, wound repair was strongly accelerated due to enhanced macrophage invasion, cell proliferation and angiogenesis.[56] A role of endogenous FGF-BP1 in wound healing also seems likely, since its expression is strongly induced upon skin wounding — in particular in keratinocytes.[57,58] However, this will have to be determined using knockout mice or neutralizing antibodies.

3. FGF Signaling in Homeostasis and Repair of the Lung Epithelium

While mice lacking FGF receptors in keratinocytes are viable and can be used to study FGF function in skin homeostasis and repair, loss of FGFR signaling in the embryonic lung epithelium caused perinatal lethality of the mice due to inhibition of lung branching morphogenesis.[45,46,59] Therefore, these mice could not be used to study the role of FGFR signaling in homeostasis and repair of the lung epithelium. A function of FGFs in the adult lung seems, however, very likely due to the strong induction of FGF7 and/or FGF10 in response to different types of lung injury (reviewed by[5]). Therefore, mice with conditional expression of a soluble FGF receptor mutant in respiratory epithelial cells of the lung were generated. The resulting inhibition of FGFR signaling did not disturb lung homeostasis, but the susceptibility of male mice to hyperoxia was enhanced and the subsequent recovery was inhibited.[60] A role of FGFR signaling in alveolar epithelial repair was also observed in studies with FGF2 knockout mice. In response to bleomycin, these mice suffered from poor recovery of epithelial integrity. Furthermore, repair of naphthalene-induced damage of the club cell secretory protein-positive bronchial epithelium was impaired.[61] Another study demonstrated that naphthalene-induced lung damage induced expression and secretion of FGF10 by parabronchial smooth muscle cells, which activated surviving Clara cells. The latter then underwent a transient epithelial to mesenchymal transition to initiate the repair process.[62] Consistent with a beneficial role of FGF10 in lung repair, inducible overexpression of this protein in the lung epithelium of transgenic mice attenuated bleomycin-induced pulmonary fibrosis.[63] In addition, mice with inducible overexpression of FGF7 in the lung epithelium were protected from hyperoxia-induced lung epithelial cell death through activation of the anti-apoptotic Akt signaling pathway.[64] Finally, FGF1 reverted the transforming growth factor β-induced epithelial-mesenchymal transition of lung epithelial cells, thereby promoting epithelial integrity in response to a pro-fibrotic stimulus.[65] Interestingly, FGF10 haploinsufficiency is associated with chronic obstructive pulmonary disease (COPD), suggesting that endogenous FGF10 is also a crucial regulator of homeostasis and repair of the human lung epithelium.[66] These studies also point to a therapeutic potential of recombinant FGFR2b ligands in pulmonary injury and disease, and several preclinical studies indeed showed therapeutic efficacy of FGF7 or FGF10 in different animal models for lung injury/disease (reviewed by[5]). By contrast, overexpression of FGFs that stimulate airway smooth muscle cells is likely to be deleterious as suggested by the enhanced expression of FGFR1 and its ligands FGF1 and FGF2 in lung biopsies of COPD patients.[67] Furthermore, increased expression of FGF1 and FGFRc isoforms was seen in patients with idiopathic pulmonary fibrosis, which most likely contributes to the pathogenesis of this disease by supporting fibroblast migration.[68] Therefore, maintenance of lung homeostasis requires a tightly regulated balance between different FGF family members.

4. FGFs in Liver Homeostasis and Repair

Several FGFs are expressed in the normal and regenerating liver as well as in the spleen from where they can reach the liver via the portal vein. This is likely to be important for liver regeneration, since a mitogenic effect of several FGFs for hepatocytes has been demonstrated *in vitro* and *in vivo*[69] (and references therein). To determine the role of FGFR signaling in liver homeostasis and regeneration, mice expressing a dominant-negative mutant of FGFR2b in hepatocytes were generated. These mice did not exhibit obvious alterations in liver homeostasis,[69] while FGFR4 knockout mice showed elevated cholesterol metabolism and bile acid synthesis.[70] When the mice expressing dominant-negative FGFR2b were subjected to partial hepatectomy, proliferation of hepatocytes was impaired due to delayed G1/S transition.[69] Impaired liver regeneration was also observed in zebrafish expressing a dominant-negative FGFR mutant in an inducible manner.[71] Since the dominant-negative receptor mutant blocks signaling through all FGF receptors in response to common ligands, the type of receptor involved in liver regeneration remains to be determined. Loss of

FGFR4 in mice aggravated carbon tetrachloride-induced liver fibrosis, but did not affect the regeneration process after partial hepatectomy,[70] suggesting that signaling by other FGF receptors compensates for the loss of FGFR4 under these conditions. However, when FGFR4 was knocked down in hepatocytes using siRNA delivered via nanoparticles, liver regeneration after partial hepatectomy was severely impaired in these mice. This was caused by enhanced apoptosis as a result of elevated levels of intrahepatic toxic bile acids. In addition, there was a severe defect in hepatocyte proliferation caused by inhibition of FGF15-FGFR4-STAT3 signaling, which is required for injury-induced expression of the FOXM1 transcription factor and subsequent cell cycle progression.[72] The discrepancy between the data obtained with FGFR4 knockout compared to knock-down mice may result from the acquisition of compensatory mechanisms during embryonic and/or postembryonic development of the total knockout mice. Consistent with an important role of FGFR4 in liver regeneration, mice lacking FGF15, a major ligand of FGFR4, showed a similar and even more severe regeneration defect as FGFR4 knock-down mice.[73,74]

Mice lacking FGFR1, FGFR2 or both receptors in hepatocytes did not display abnormalities under unchallenged conditions and their response to carbon tetrachloride-induced liver injury and fibrosis was not affected. However, severe hepatocyte necrosis occurred after partial hepatectomy, which resulted from impaired expression of the DBP and TEF transcription factors and their target genes, which encode important enzymes involved in compound detoxification. Therefore, the liver tissue that remained after partial hepatectomy (one third of the liver) failed to metabolize endogenous compounds and the drugs applied for anaesthesia/analgesia.[75] Consistent with an important role of FGFR2b in liver regeneration, FGF7 expression was induced in mice and in patients with major liver injury, and FGF7 knockout mice had a severe reduction in the expansion of liver progenitor cells and a higher mortality after toxin-induced injury. Vice versa, overexpression of FGF7 in the liver of transgenic mice induced the expansion of these progenitor cells and ameliorated hepatic dysfunction.[76] In addition to ligands of

FGFR1b and FGFR2b, expression of FGF9 was induced after carbon tetrachloride-induced liver injury. This upregulation occurred in hepatic stellate cells and resulted in enhanced proliferation of hepatocytes,[77] possibly via activation of FGFR4. This finding suggests that several FGFs and FGF receptors contribute to efficient liver regeneration. Indeed, when FGFR4 was knocked-down in hepatocytes of mice lacking FGFR1 and FGFR2, liver failure occurred within 2–3 days after partial hepatectomy, demonstrating that FGFR signaling is essential for liver regeneration under these conditions.[72]

5. FGF Signaling in Intestinal Homeostasis and Repair

Recent studies demonstrated that FGF signaling also controls intestinal development and homeostasis. Thus, mice lacking FGF9, which is produced by intestinal epithelial cells, or mice lacking the mesenchymal receptors for FGF9, were characterized by a shortened small intestine, demonstrating that FGF signaling regulates elongation of this part of the gut.[78] FGF9 is also essential for the development of the cecum. It was shown to signal to mesenchymal FGF receptors during embryonic development, which in turn induces expression of FGF10 in the mesenchyme and consequent stimulation of FGF receptors on epithelial cells. This double paracrine loop is essential for epithelial budding and subsequent cecum development.[79,80] Gain- and loss-of-function experiments in the adult mouse intestine demonstrated that FGF10 increases the number of Goblet cells on the expense of Paneth cells, thereby altering the balance between these cell types in the adult mouse small intestine.[81] A role of FGFR3 in intestinal homeostasis was observed by analysis of FGFR3 knockout mice. These mice showed increased intestinal crypt depth due to increased intestinal crypt cell proliferation, whereas the villae length or the distribution of differentiated intestinal cells were not affected. The stimulatory effect of a loss of FGFR3 signaling on crypt cell proliferation was confirmed in wild-type mice upon treatment with anti-FGFR3 antibodies. *Vice versa,* injection of the FGFR3 ligand FGF18 reduced cell proliferation in

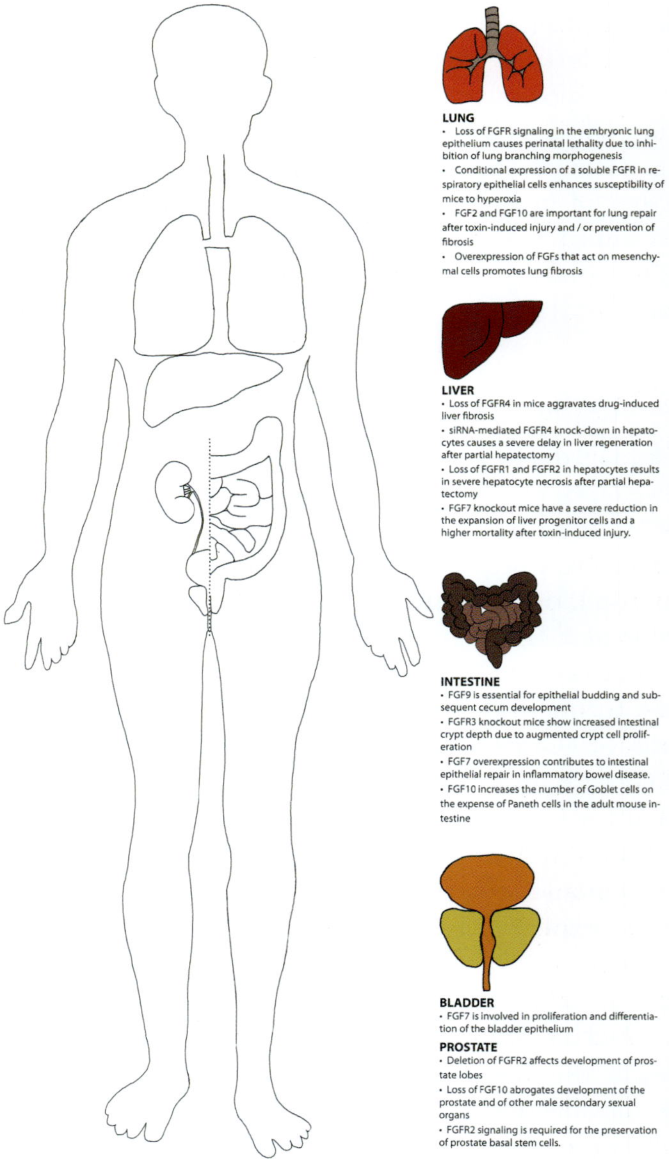

Fig. 2. Functions of FGFs in homeostasis and repair of selected organs. The organs, for which FGF functions are described in this article, are shown schematically together with the functions of FGFs in homeostasis and repair of these organs.

the crypts. These results identify FGFR3 and its ligands as negative regulators of the intestinal crypt, which include the progenitor cells.[82]

A role of FGF signaling in intestinal repair was suggested by the finding that FGF7 is strongly overexpressed in inflammatory bowel disease, a chronic inflammatory disease that is characterized by severe tissue damage in the intestine.[83–85] This upregulation most likely prevents a more severe injury and contributes to intestinal epithelial repair, since FGF7 knockout mice as well as mice lacking FGF7-producing intestinal γδ intraepithelial T lymphocytes were more sensitive to dextran sodium sulfate induced mucosal injury than wild-type mice. Furthermore, these animals exhibited delayed repair of the tissue after termination of the treatment.[86]

6. FGF Signaling in Prostate and Bladder Homeostasis

Most studies on FGFs in the prostate addressed the roles of these mitogens and their receptors in cancer development and progression (reviewed by[87]). However, FGFs also control prostate development and homeostasis. Thus, deletion of FGFR2 in prostatic epithelial precursor cells of mice strongly affected development of the anterior and ventral prostate lobes and disrupted tissue development as well as the responsiveness of the dorsolateral prostate to androgens.[88] This seems to be predominantly mediated via FGF10, since disruption of the *FGF10* gene in mice abrogated development of the prostate and of other male secondary sexual organs, whereas FGF7-deficient mice have no obvious defects in the male reproductive tract.[19,89] In addition, FGFR2 signaling was shown to be required for the preservation of stemness and for prevention of differentiation of prostate basal stem cells.[90] It remains to be determined if FGFs also control the response of the prostate to tissue injury.

A role of FGF7 in epithelial homeostasis of the bladder was demonstrated by the impaired proliferation and differentiation of the bladder epithelium in FGF7 knockout mice.[91] Repair of this organ upon partial outlet obstruction was, however, not impaired in these mice, most likely due to compensatory upregulation of FGF10.[92]

7. Conclusions and Outlook

In recent years, FGFs have emerged as crucial regulators of epithelial homeostasis and repair. FGF function in these processes has been studied in most detail in the organs covered in this article, but they are obviously not restricted to these tissues. Indeed, FGFR signaling controls embryonic development of additional tissues and organs, and in particular FGF7 is upregulated in response to injury of various epithelial organs (reviewed by[5,93]), and it stimulates proliferation, migration, and survival of different types of epithelial cells (reviewed by[94]). Importantly, recombinant FGFs have a remarkable therapeutic potential for the restoration of epithelial homeostasis after various types of insults. Thus, FGF7 is already therapeutically approved for the treatment of mucositis, a severe inflammation of the oral mucosa and the gastrointestinal tract that occurs in response to chemo- and radiotherapy,[5,95] and preclinical studies suggest that it may also have a beneficial effect in other diseases associated with epithelial injury, such as graft-versus-host disease and various types of pulmonary diseases.[5] Therefore, future basic and clinical research in the FGF field will unravel novel and exciting functions and clinical applications of FGFs in epithelial homeostasis and repair.

Acknowledgements

Research on FGFs in our laboratory is supported by the Swiss National Science Foundation (310030_132884 to S.W) and the ETH Zurich (to S.W.).

References

1. Beenken, A. and Mohammadi, M. The FGF family: Biology, pathophysiology and therapy. *Nat Rev Drug Discov* **8**, 235–253 (2009).
2. Ornitz, D.M. and Itoh, N. The Fibroblast Growth Factor signaling pathway. *Wiley Interdiscip Rev Dev Biol* **4**, 215–266 (2015).
3. Grose, R. and Dickson, C. Fibroblast growth factor signaling in tumorigenesis. *Cytokine Growth Factor Rev* **16**, 179–286 (2005).
4. Turner, N. and Grose, R. Fibroblast growth factor signalling: From development to cancer. *Nat Rev Cancer* **10**, 116–129 (2010).

5. Finch, P.W., Mark Cross, L.J., McAuley, D.F. and Farrell, C.L. Palifermin for the protection and regeneration of epithelial tissues following injury: New findings in basic research and pre-clinical models. *J Cell Mol Med* **17**, 1065–1687 (2013).
6. Zhang, X., *et al.* Receptor specificity of the fibroblast growth factor family. The complete mammalian FGF family. *J Biol Chem* **281**, 15694–15700 (2006).
7. Finch, P.W. and Rubin, J.S. Keratinocyte growth factor/fibroblast growth factor 7, a homeostatic factor with therapeutic potential for epithelial protection and repair. *Adv Cancer Res* **91**, 69–136 (2004).
8. Zhang, J. and Li, Y. Therapeutic uses of FGFs. *Semin Cell Dev Biol* (2015).
9. Werner, S., *et al.* Targeted expression of a dominant-negative FGF receptor mutant in the epidermis of transgenic mice reveals a role of FGF in keratinocyte organization and differentiation. *EMBO J* **12**, 2635–2643 (1993).
10. Yang, J., *et al.* Fibroblast growth factor receptors 1 and 2 in keratinocytes control the epidermal barrier and cutaneous homeostasis. *J Cell Biol* **188**, 935–952 (2010).
11. Ferone, G., *et al.* Mutant p63 causes defective expansion of ectodermal progenitor cells and impaired FGF signalling in AEC syndrome. *EMBO Mol Med* **4**, 192–205 (2012).
12. Meyers, G.A., Orlow, S.J., Munro, I.R., Przylepa, K.A., *et al.* Fibroblast growth factor receptor 3 (FGFR3) transmembrane mutation in Crouzon syndrome with acanthosis nigricans. *Nat Genet* **11**, 462–464 (1995).
13. Hafner, C., *et al.* Mosaicism of activating FGFR3 mutations in human skin causes epidermal nevi. *J Clin Invest* **116**, 2201–2207 (2006).
14. Logie, A., *et al.* Activating mutations of the tyrosine kinase receptor FGFR3 are associated with benign skin tumors in mice and humans. *Hum Mol Genet* **14**, 1153–1160 (2005).
15. Mandinova, A., *et al.* A positive FGFR3/FOXN1 feedback loop underlies benign skin keratosis versus squamous cell carcinoma formation in humans. *J Clin Invest* **119**, 3127–3137 (2009).
16. Petiot, A., *et al.* A crucial role for FGFR2-IIIb signalling in epidermal development and hair follicle patterning. *Development* **130**, 5493–5501 (2003).
17. Werner, S., *et al.* The function of KGF in morphogenesis of epithelium and reepithelialization of wounds. *Science* **266**, 819–822 (1994).
18. Suzuki, K., *et al.* Defective terminal differentiation and hypoplasia of the epidermis in mice lacking the FGF10 gene. *FEBS Lett* **481**, 53–56 (2000).

19. Guo, L., Degenstein, L. and Fuchs, E. Keratinocyte growth factor is required for hair development but not for wound healing. *Genes Dev* **10**, 165–175 (1996).
20. Grose, R., *et al.* The role of fibroblast growth factor receptor 2b in skin homeostasis and cancer development. *EMBO J* **26**, 1268–1278 (2007).
21. Rosenquist, T.A. and Martin, G.R. Fibroblast growth factor signalling in the hair growth cycle: Expression of the fibroblast growth factor receptor and ligand genes in the murine hair follicle. *Dev Dyn* **205**, 379–386 (1996).
22. Greco, V., *et al.* A two-step mechanism for stem cell activation during hair regeneration. *Cell Stem Cell* **4**, 155–169 (2009).
23. Kimura-ueki, M., *et al.* Hair cycle resting phase is regulated by cyclic epithelial FGF18 signaling. *J Invest Dermatol* **132**, 1338–1345 (2012).
24. Leishman, E., *et al.* Foxp1 maintains hair follicle stem cell quiescence through regulation of FGF18. *Development* **140**, 3809–3818 (2013).
25. Zhang, K., *et al.* Fibroblast growth factor-peptide improves barrier function and proliferation in human keratinocytes after radiation. *Int J Radiat Oncol Biol Phys* **81**, 248–254 (2011).
26. Sulcova, J., *et al.* Mast cells are dispensable in a genetic mouse model of chronic dermatitis. *Am J Pathol* **185**, 1575–1587 (2015).
27. Sulcova, J., Maddaluno, L., Meyer, M. and Werner, S. Accumulation and activation of epidermal gammadelta T cells in a mouse model of chronic dermatitis is not required for the inflammatory phenotype. *Eur J Immunol* **45**, 2517–2528 (2015).
28. Guttman-Yassky, E., *et al.* Broad defects in epidermal cornification in atopic dermatitis identified through genomic analysis. *J Allergy Clin Immunol* **124**, 1235–1244 e58 (2009).
29. Finch, P.W., Murphy, F., Cardinale, I. and Krueger, J.G. Altered expression of keratinocyte growth factor and its receptor in psoriasis. *Am J Pathol* **151**, 1619–1628 (1997).
30. Kovacs, D., *et al.* Immunohistochemical analysis of keratinocyte growth factor and fibroblast growth factor 10 expression in psoriasis. *Exp Dermatol* **14**, 130–137 (2005).
31. Guo, L., Yu, Q.C. and Fuchs, E. Targeting expression of keratinocyte growth factor to keratinocytes elicits striking changes in epithelial differentiation in transgenic mice. *EMBO J* **12**, 973–986 (1993).
32. Werner, S., *et al.* Large induction of keratinocyte growth factor expression in the dermis during wound healing. *Proc Natl Acad Sci USA* **89**, 6896–6900 (1992).

33. Marchese, C., *et al.* Modulation of keratinocyte growth factor and its receptor in reepithelializing human skin. *J Exp Med* **182**, 1369–1376 (1995).
34. Brauchle, M., Angermeyer, K., Hubner, G. and Werner, S. Large induction of keratinocyte growth factor expression by serum growth factors and pro-inflammatory cytokines in cultured fibroblasts. *Oncogene* **9**, 3199–3204 (1994).
35. Chedid, M., Rubin, J.S., Csaky, K.G. and Aaronson, S.A. Regulation of keratinocyte growth factor gene expression by interleukin 1. *J Biol Chem* **269**, 10753–10757 (1994).
36. Boismenu, R. and Havran, W.L. Modulation of epithelial cell growth by intraepithelial gamma delta T cells. *Science* **266**, 1253–1255 (1994).
37. Jameson, J., *et al.* A role for skin gammadelta T cells in wound repair. *Science* **296**, 747–749 (2002).
38. Werner, S., Breeden, M., Hubner, G., Greenhalgh, D.G., *et al.* Induction of keratinocyte growth factor expression is reduced and delayed during wound healing in the genetically diabetic mouse. *J Invest Dermatol* **103**, 469–473 (1994).
39. Brauchle, M., Fassler, R. and Werner, S. Suppression of keratinocyte growth factor expression by glucocorticoids *in vitro* and during wound healing. *J Invest Dermatol* **105**, 579–584 (1995).
40. Peng, C., *et al.* Lack of FGF-7 further delays cutaneous wound healing in diabetic mice. *Plast Reconstr Surg* **128**, 673e–684e (2011).
41. Beer, H.D., *et al.* Mouse fibroblast growth factor 10: cDNA cloning, protein characterization, and regulation of mRNA expression. *Oncogene* **15**, 2211–2218 (1997).
42. Nakatake, Y., Hoshikawa, M., Asaki, T., Kassai, Y., *et al.* Identification of a novel fibroblast growth factor, FGF-22, preferentially expressed in the inner root sheath of the hair follicle. *Biochim Biophys Acta* **1517**, 460–463 (2001).
43. Beyer, T.A., Werner, S., Dickson, C. and Grose, R. Fibroblast growth factor 22 and its potential role during skin development and repair. *Exp Cell Res* **287**, 228–236 (2003).
44. Komi-Kuramochi, A., *et al.* Expression of fibroblast growth factors and their receptors during full-thickness skin wound healing in young and aged mice. *J Endocrinol* **186**, 273–289 (2005).
45. Sekine, K., *et al.* Fgf10 is essential for limb and lung formation. *Nat Genet* **21**, 138–141 (1999).
46. Min, H., *et al.* Fgf-10 is required for both limb and lung development and exhibits striking functional similarity to Drosophila branchless. *Genes Dev* **12**, 3156–3161 (1998).

47. Jarosz, M., *et al.* Fibroblast growth factor 22 is not essential for skin development and repair but plays a role in tumorigenesis. *PLoS One* **7**, e39436 (2012).
48. Meyer, M., *et al.* FGF receptors 1 and 2 are key regulators of keratinocyte migration *in vitro* and in wounded skin. *J Cell Sci* **125**, 5690–5701 (2012).
49. Fuhr, M.J., *et al.* A modeling approach to study the effect of cell polarization on keratinocyte migration. *PLoS One* **10**, e0117676 (2015).
50. Usui, M.L., Mansbridge, J.N., Carter, W.G., Fujita, M., *et al.* Keratinocyte migration, proliferation, and differentiation in chronic ulcers from patients with diabetes and normal wounds. *J Histochem Cytochem* **56**, 687–696 (2008).
51. Gay, D., *et al.* FGF9 from dermal gammadelta T cells induces hair follicle neogenesis after wounding. *Nat Med* **19**, 916–923 (2013).
52. Broadley, K.N., *et al.* Monospecific antibodies implicate basic fibroblast growth factor in normal wound repair. *Lab Invest* **61**, 571–575 (1989).
53. Ortega, S., Ittmann, M., Tsang, S.H., Ehrlich, M., *et al.* Neuronal defects and delayed wound healing in mice lacking fibroblast growth factor 2. *Proc Natl Acad Sci USA* **95**, 5672–5677 (1998).
54. Miller, D.L., Ortega, S., Bashayan, O., Basch, R. *et al.* Compensation by fibroblast growth factor 1 (FGF1) does not account for the mild phenotypic defects observed in FGF2 null mice. *Mol Cell Biol* **20**, 2260–2268 (2000).
55. Oladipupo, S.S., *et al.* Endothelial cell FGF signaling is required for injury response but not for vascular homeostasis. *Proc Natl Acad Sci USA* **111**, 13379–13384 (2014).
56. Tassi, E., *et al.* Impact of fibroblast growth factor-binding protein-1 expression on angiogenesis and wound healing. *Am J Pathol* **179**, 2220–2232 (2011).
57. Kurtz, A., *et al.* Differential regulation of a fibroblast growth factor-binding protein during skin carcinogenesis and wound healing. *Neoplasia* **6**, 595–602 (2004).
58. Beer, H.D., *et al.* The fibroblast growth factor binding protein is a novel interaction partner of FGF-7, FGF-10 and FGF-22 and regulates FGF activity: Implications for epithelial repair. *Oncogene* **24**, 5269–5277 (2005).
59. Peters, K., *et al.* Targeted expression of a dominant negative FGF receptor blocks branching morphogenesis and epithelial differentiation of the mouse lung. *EMBO J* **13**, 3296–3301 (1994).

60. Hokuto, I., Perl, A.K. and Whitsett, J.A. FGF signaling is required for pulmonary homeostasis following hyperoxia. *Am J Physiol Lung Cell Mol Physiol* **286**, L580–L587 (2004).
61. Guzy, R.D., Stoilov, I., Elton, T.J., Mecham, R.P., *et al.* Fibroblast growth factor 2 is required for epithelial recovery, but not for pulmonary fibrosis, in response to bleomycin. *Am J Respir Cell Mol Biol* **52**, 116–128 (2015).
62. Volckaert, T., *et al.* Parabronchial smooth muscle constitutes an airway epithelial stem cell niche in the mouse lung after injury. *J Clin Invest* **121**, 4409–4419 (2011).
63. Gupte, V.V., *et al.* Overexpression of fibroblast growth factor-10 during both inflammatory and fibrotic phases attenuates bleomycin-induced pulmonary fibrosis in mice. *Am J Respir Crit Care Med* **180**, 424–436 (2009).
64. Ray, P., *et al.* Inducible expression of keratinocyte growth factor (KGF) in mice inhibits lung epithelial cell death induced by hyperoxia. *Proc Natl Acad Sci USA* **100**, 6098–6103 (2003).
65. Ramos, C., *et al.* FGF-1 reverts epithelial-mesenchymal transition induced by TGF-{beta}1 through MAPK/ERK kinase pathway. *Am J Physiol Lung Cell Mol Physiol* **299**, L222–L331 (2010).
66. Klar, J., *et al.* Fibroblast growth factor 10 haploinsufficiency causes chronic obstructive pulmonary disease. *J Med Genet* **48**, 705–709 (2011).
67. Kranenburg, A.R., *et al.* Chronic obstructive pulmonary disease is associated with enhanced bronchial expression of FGF-1, FGF-2, and FGFR-1. *J Pathol* **206**, 28–38 (2005).
68. MacKenzie, B., *et al.* Increased FGF1-FGFRc expression in idiopathic pulmonary fibrosis. *Respir Res* **16**, 83 (2015).
69. Steiling, H., *et al.* Fibroblast growth factor receptor signalling is crucial for liver homeostasis and regeneration. *Oncogene* **22**, 4380–4388 (2003).
70. Yu, C., *et al.* Elevated cholesterol metabolism and bile acid synthesis in mice lacking membrane tyrosine kinase receptor FGFR4. *J Biol Chem* **275**, 15482–15489 (2000).
71. Kan, N.G., Junghans, D. and Izpisua Belmonte, J.C. Compensatory growth mechanisms regulated by BMP and FGF signaling mediate liver regeneration in zebrafish after partial hepatectomy. *FASEB J* **23**, 3516–3525 (2009).
72. Padrissa-Altes, S., *et al.* Control of hepatocyte proliferation and survival by FGF receptors is essential for liver regeneration in mice. *Gut* **64**, 1444–1453 (2015).

73. Uriarte, I., *et al.* Identification of fibroblast growth factor 15 as a novel mediator of liver regeneration and its application in the prevention of post-resection liver failure in mice. *Gut* **62**, 899–910 (2013).
74. Kong, B., *et al.* Fibroblast growth factor 15 deficiency impairs liver regeneration in mice. *Am J Physiol Gastrointest Liver Physiol* **306**, G893–G902 (2014).
75. Bohm, F., *et al.* FGF receptors 1 and 2 control chemically induced injury and compound detoxification in regenerating livers of mice. *Gastroenterology* **139**, 1385–1396 (2010).
76. Takase, H.M., *et al.* FGF7 is a functional niche signal required for stimulation of adult liver progenitor cells that support liver regeneration. *Genes Dev* **27**, 169–181 (2013).
77. Antoine, M., *et al.* Expression and function of fibroblast growth factor (FGF) 9 in hepatic stellate cells and its role in toxic liver injury. *Biochem Biophys Res Commun* **361**, 335–341 (2007).
78. Geske, M.J., Zhang, X., Patel, K.K., Ornitz, D.M., *et al.*. FGF9 signaling regulates small intestinal elongation and mesenchymal development. *Development* **135**, 2959–2968 (2008).
79. Zhang, X., *et al.* Reciprocal epithelial-mesenchymal FGF signaling is required for cecal development. *Development* **133**, 173–180 (2006).
80. Al Alam, D., *et al.* FGF9-Pitx2-FGF10 signaling controls cecal formation in mice. *Dev Biol* **369**, 340–348 (2012).
81. Al Alam, D., *et al.* Fibroblast growth factor 10 alters the balance between goblet and Paneth cells in the adult mouse small intestine. *Am J Physiol Gastrointest Liver Physiol* **308**, G678–G690 (2015).
82. Arnaud-Dabernat, S., Yadav, D. and Sarvetnick, N. FGFR3 contributes to intestinal crypt cell growth arrest. *J Cell Physiol* **216**, 261–268 (2008).
83. Bajaj-Elliott, M., Breese, E., Poulsom, R., Fairclough, P.D., *et al.* Keratinocyte growth factor in inflammatory bowel disease. Increased mRNA transcripts in ulcerative colitis compared with Crohn's disease in biopsies and isolated mucosal myofibroblasts. *Am J Pathol* **151**, 1469–1476 (1997).
84. Finch, P.W., Pricolo, V., Wu, A. and Finkelstein, S.D. Increased expression of keratinocyte growth factor messenger RNA associated with inflammatory bowel disease. *Gastroenterology* **110**, 441–451 (1996).
85. Brauchle, M., *et al.* Keratinocyte growth factor is highly overexpressed in inflammatory bowel disease. *Am J Pathol* **149**, 521–529 (1996).

86. Chen, Y., Chou, K., Fuchs, E., Havran, W.L. *et al.* Protection of the intestinal mucosa by intraepithelial gamma delta T cells. *Proc Natl Acad Sci USA* **99**, 14338–14343 (2002).
87. Lin, Y. and Wang, F. FGF signalling in prostate development, tissue homoeostasis and tumorigenesis. *Biosci Rep* **30**, 285–291 (2010).
88. Lin, Y., *et al.* Fibroblast growth factor receptor 2 tyrosine kinase is required for prostatic morphogenesis and the acquisition of strict androgen dependency for adult tissue homeostasis. *Development* **134**, 723–734 (2007).
89. Donjacour, A.A., Thomson, A.A. and Cunha, G.R. FGF-10 plays an essential role in the growth of the fetal prostate. *Dev Biol* **261**, 39–54 (2003).
90. Huang, Y., *et al.* Type 2 Fibroblast Growth Factor Receptor Signaling Preserves Stemness and Prevents Differentiation of Prostate Stem Cells from the Basal Compartment. *J Biol Chem* **290**, 17753–17761 (2015).
91. Tash, J.A., David, S.G., Vaughan, E.E. and Herzlinger, D.A. Fibroblast growth factor-7 regulates stratification of the bladder urothelium. *J Urol* **166**, 2536–2541 (2001).
92. Lendvay, T.S., *et al.* Compensatory paracrine mechanisms that define the urothelial response to injury in partial bladder outlet obstruction. *Am J Physiol Renal Physiol* **293**, F1147–F1156 (2007).
93. Werner, S. Keratinocyte growth factor: A unique player in epithelial repair processes. *Cytokine Growth Factor Rev* **9**, 153–165 (1998).
94. Braun, S., auf dem Keller, U., Steiling, H. and Werner, S. Fibroblast growth factors in epithelial repair and cytoprotection. *Philos Trans R Soc Lond B Biol Sci* **359**, 753–757 (2004).
95. Vadhan-Raj, S., Goldberg, J.D., Perales, M.A., Berger, D.P., *et al.* Clinical applications of palifermin: Amelioration of oral mucositis and other potential indications. *J Cell Mol Med* **17**, 1371–1384 (2013).

Chapter 9
Growth Factors in the Eye

Rong Ju, Chunsik Lee, Weisi Lu and Xuri Li*

Abstract

The eye is an important part of the central nervous system and has pivotal roles in health and life. Growth factors are critically required for the development and function of the eye. Among different growth factors, the fibroblast growth factors (FGF2) and platelet-derived growth factors (PDGFs) are of particular importance due to their critical function in regulating multiple processes of eye development and physiology. Indeed, the FGFs, PDGFs and their receptors are highly expressed in normal and diseased eyes. Moreover, their expressions are tightly-controlled and finely regulated to ensure normal eye development and function. On the other hand, dysregulation of the FGFs and PDGFs can lead to various ocular diseases. In this chapter, we discuss the expression profiles of the FGFs, PDGFs and their receptors as well as their physiological roles in eye development, function, and diseases. A better understanding of the expression, regulation and working mechanisms of these important growth factors may facilitate the possibilities of better therapeutics for eye-related diseases.

State Key Laboratory of Ophthalmology, Zhongshan Ophthalmic Center, Sun Yat-Sen University, 54 South Xianlie Road, Guangzhou 510060, Guangdong, P. R. China.

* Corresponding author: E-mail: lixr6@mail.sysu.edu.cn

I. FGFs in the Eye

1.1. *Expression of FGFs and FGFRs in the eye*

The fibroblast growth factor (FGF) family contains 22 members in mammals that are structurally and functionally related. The prototype FGFs are the acidic FGF (FGF-1) and basic FGF (FGF-2), which binds to five FGF receptors (FGFRs).[1] As found in most types of cells, the FGFs and their receptors are highly expressed in different eye tissues. *In situ* hybridization shows that in adult rats, FGF1 is produced by multiple sources, including photoreceptor visual cells, neuronal cells in the inner nuclear layer, ganglion cells of the retina, pigment epithelial cells, iris and ciliary body, epithelial cells of the cornea, conjunctiva and lens. On the other hand, FGF2 is synthesized mainly by the photoreceptor visual cells.[2] FGFR1 and FGFR2 are identified widely in the adult central nervous system, including the retina.[3] FGFR4, however, is found to be expressed only in adult retinal neurons, being most prominent in the outer nuclear layer.[4]

During mouse development, FGF1 and FGF2 are expressed in the neural retina and lens cells.[5–7] FGF1 is detected in almost all the ocular tissues with the highest expression in the retina, corneal and vitreous body. FGFR1-4 are expressed in lens cells while FGFR1 and FGFR2 are primarily found during retinal development.[3,8,9] FGF2 mRNA expression is not detected in the retina until postnatal day 10 (P10) and maintains a similar level in adult in mice.

A variety of pathological conditions and stresses regulate FGF/FGFR expression in eye tissues. Thus, the expression of FGFR1 is upregulated by light stress in the outer nuclear layer (ONL) and outer plexiform layer (OPL), and it is co-localizes with FGF2.[10] FGF1 is detected in the pre-retinal membranes in the setting of proliferative vitreoretinopathy (PVR)[11] and in diabetic proliferative retinopathy.[12]

Mechanical injury induces FGF2 and FGFR1 expression in both mouse and rat retinae, although the injury-induced upregulation of FGF2 is much less pronounced in the mouse compared to the rat retina.[13] FGFR1 in the rat retina is concentrated primarily in the perinuclear cytoplasm of photoreceptor cell bodies. Experimental retinal detachment or focal injury triggers a rapid increase in

FGFR1 levels in the outer nuclear layer that persists for at least 7 days.[14] In the retinae of rd (C57BL/6J rd/rd) mice, a naturally occurring mutant strain with retinal neuron degeneration, elevated expression of FGF2 in the degenerating neurons is detected in the outer nuclear layer during photoreceptor degeneration.[15] Gene expression of FGF1 and FGFR1 is detected in the ganglion cells and inner nuclear layers in normal adult rats. Laser photocoagulation upregulates the expressions of FGF2 and FGFR1 in proliferating retinal pigment epithelial (RPE) cells. In addition, macrophage-like cells that migrate into the lesion, also display FGF2 expression.[16] In cultured RPE cells *in vitro*, FGF2 mRNA level is increased by constant light, brain-derived neurotrophic factor (BDNF), ciliary neurotrophic factor, interleukin-1 beta and oxidizing agents.[16] In postnatal day 16–24 rats, hyperoxia slows photoreceptor death in a dose-related fashion with decreased FGF2 protein levels, whereas hypoxia accelerated cell death with increased FGF2 levels.[17] In oxygen-induced retinopathy, the expression of FGF2 was downregulated by hyperoxia and upregulated by hypoxia.[18]

1.2. *FGFs and FGFRs in ocular neovascular diseases*

Ocular angiogenesis is a devastating pathology of many diseases leading to patients' blindness, such as retinopathy of prematurity (ROP), proliferative diabetic retinopathy (DR), wet age-related macular degeneration (AMD), retina vein occlusion, and corneal neovascularization.[19] ROP and DR are the leading causes of blindness in children and working population globally,[20,21] while AMD is the major cause of vision loss in the elderly population in developed countries.

AMD affects about eleven million people in the United States alone.[22] ROP causes blindness in premature children, accounting for 6–18% of total childhood blindness.[23] In ROP, oxygen therapy during neonatal intensive care causes regression of retinal vasculature. When returned to normoxia, the retinae suffer from severe ischemia, which subsequently upregulates VEGF expression, thus leading to retinal neovascularization.[24] In DR and retina vein occlusion, activated retinal neovessels grow into the vitreous. These rapidly growing vessels are highly leaky and prone to hemorrhage. This can lead to the collapse of the vitreous and detachment of the retina, resulting in impaired vision.[25]

In the case of AMD, pathological new vessels grow from the choroid and extend through the Bruch's membrane and retinal pigment epithelium (RPE), causing detachment of photoreceptors from the RPE and, consequently, the death of photoreceptors and ultimately blindness.[26] Corneal neovascularization is another vision-threatening conditions involving pathological angiogenesis. The causes of corneal neovascularization are multiple, including pathogen infections, inflammation on the ocular surface, and trauma. Vascularization in the cornea leads to scarring, edema, and greater risk of graft rejection.[25,27]

The FGF/FGFR pathway has been shown to play an important role in ocular vascularization. Two prototypic FGFs, FGF2 (bFGF) and FGF1 (aFGF) have been studied extensively. FGF2 is a potent angiogenic molecule *in vitro* and *in vivo*.[28] It increases proliferation and migration of endothelial cells and promotes the production of proteinase and expression of specific integrins in cultured endothelial cells.[29,30] FGF2 is known to exert its angiogenic effect via VEGF-dependent as well as VEGF-independent pathways.[31,32] One study compared the antiangiogenic effects of bevacizumab alone and after its combination with anti-bFGF antibodies. It showed that EC sprouting cannot be fully inhibited by selective inhibition of either VEGF or bFGF. However, a combined inhibition of both signaling pathways markedly reduced EC sprouting.[33] It is important to note, however, that even combined inhibition did not fully reduce EC sprouting to the baseline level, and that genetic ablation of FGF2 did not inhibit the formation of laser-induced CNV.[34] FGF2 induces the secretion of VEGF and HGF by Müller glial cells and stimulates cell proliferation.

More extensive neovascularization involves integrated VEGF and FGF signaling. For example, maximum co-labelling of VEGF and FGF2 in rat RPE and choriocapillaris cells occurs in CNV between 2 and 7 days after laser treatment. In human studies, Amin *et al* has shown that FGF1 and FGF2 are upregulated in the choroidal neovascular membranes of AMD patients in RPE cells and endothelial cells. Thus, these studies suggest important roles of FGFs in CNV of AMD.[35]

FGF2 is also important for the survival and maturation of neurons and glial cells and plays an important role in tissue repair.[36] The concentration of FGF2 in the vitreous is increased in patients with proliferative diabetic retinopathy (PDR).[37] FGF2 is therefore involved in the pathogenesis of PDR[38] and stimulates VEGF production. More importantly, inhibition of VEGF resulted in elevated level of FGF2. Yet another study suggests that the increased VEGF-A level in PDR might be responsible for the impairment of retinal function, whereas the upregulation of FGF2 might be a compensatory mechanism for neuroprotection and vasculoprotection.[39]

FGF2 is involved in the formation of epiretinal membranes.[40] It might coordinate with nerve growth factor and glial cell line-derived neurotrophic factor to promote the formation of epiretinal membranes in PDR,[38] since glial cell line-derived neurotrophic factor stimulates Müller cells to produce FGF2.[41] This in turn stimulates endothelial cell proliferation and production of VEGF.[31] In addition to FGF-2, FGFRs,[40] FGF-1[12] and FGF-5[42] were all detected in the epiretinal membranes. Taken together, these studies indicate important roles of the FGF family in the pathogenesis of PDR.

FGF2 has shown a potent capacity in inducing angiogenesis in animal models of corneal neovascularization, partially by upregulation of VEGF.[43] Indeed, Cao *et al.* has reported that FGF2-induced blood vessels have more endothelial fenestrations and higher permeability compared to those induced by VEGF-A in a mouse corneal vascularization model.[44] Other FGFs also play important roles in ocular angiogenesis. FGF7, also called keratinaocyte growth factor (KGF), is reported to be involved in neovascularization in rat cornea.[45]

Mice with genetic deletion of FGF2 show defective vascular tone[46] but no obvious difference in laser-induced choroidal neovascularization[34] and retinal blood vessel development.[47] Double deletion of FGF1 and FGF2 failed to add any severity in a wound-healing model,[48] suggesting that FGF1 and FGF2 are unable to compensate for each other. Furthermore, deletion of FGFR1 resulted in embryonic lethality prior to blood vessel development.[49,50] No vascular obvious defects have been reported in FGFR2, FGFR3 or FGFR4 deficient mice.[51–53]

1.3. *FGFs and FGFRs in lens related diseases*

The growth of cataracts is a major ocular disease associated with lens defect. It is one of the biggest global health issues, particularly in developing countries. Cataracts cause vision impair in ~80 million people and blindness in ~18 million people in the world.[54] Currently, surgery is the only way to cure vision defect caused by cataracts.[55] A common ophthalmic procedure is to remove the opaque fiber mass followed by implantation of a synthetic intraocular lens (IOL).[56] Despite the effectiveness of this approach, certain surgery-related complications remain, the most frequent one being posterior capsular opacification (PCO), also termed a secondary cataract.[57] PCO is regarded to be induced by a wound-healing response in the lens triggered by surgery. Residual lens epithelial cells (LECs) on the lens capsule proliferate and migrate across the posterior capsule and undergo trans-differentiation into myofibroblasts together with lens fiber regeneration, leading to lens opacification.[57]

Members of the FGF family are also well known to play a critical role in the development and maintenance of the structure of the lens as well as differentiation of lens cells.[58] The lens is composed of a spheroidal mass of well aligned and elongated lens fiber cells, and is covered anteriorly by a monolayer of epithelial cells. During eye development, the lens derives from the optic vesicles and become lens vesicle. Posterior lens vesicle cells elongate to become the primary lens fibers whereas anterior vesicle cells differentiate into a layer of lens epithelial cells. The epithelial cells, particularly in a region above the lens equator, known as the germinative zone, continue to proliferate, migrate below the equator and differentiate into secondary fiber cells.

FGF gradient has been proposed to be the primary cue for the polarity formation of the lens. The anterior-posterior gradient of FGFs in the mammalian eye ensures the growth and polarity of the lens. Interestingly, a high concentration of FGF induces cell migration and fiber cell differentiation, whereas a low dose stimulates cell proliferation. Consistently, higher concentration of FGF is found in the vitreous than in aqueous tissue. This fits well with the notion that vitreous humor, which incubates lens fiber cells, induces lens fiber differentiation.[59] The FGF gradient has been proposed to be the key cue for lens polarity maintained in all vertebrates.[21,60–63] FGFs also play a central role in the proliferation of lens epithelial cells.[64] Given the important function of FGFs in lens epithelial cell proliferation, it is not surprising that FGFs play a role in PCO. Mansfield reported that FGFs promote the survival of cells undergoing EMT mediated by TGF-β, PCO-like ECM production and plaque formation, thus inducing PCO development after cataract surgery.[65] Another study showed that FGF2 is an autocrine factor in capsular bags and supports the survival of LECs probably via FGFR1.[66] Thus, FGFs may be important therapeutic targets for PCO.

1.4. *FGFs and FGFRs in other ocular diseases*

Many ocular diseases involve retinal degeneration, when uncontrolled, leading to vision loss and blindness. Neurotrophic factors have been found to be neuroprotective, among which are multiple FGF family members. FGF2 is reported to rescue retinal degeneration in RCS rats with inherited retinal dystrophy[67] or light damage[68] after protein treatment or gene delivery by AAV-FGF-2.[69] FGF5 and FGF18 gene delivery by AAV also decreased apoptosis of photoreceptors in transgenic rat models of retinitis pigmentosa.[70] FGF19 has also been shown to be neuroprotective on adult mammalian photoreceptors *in vitro*.[71]

Corneal epithelial cells form the first defense barrier of the eye. Trauma, surgery and other diseases can cause corneal wounds. Corneal wound healing poses special concerns because it requires the maintenance of corneal transparency and the refractive properties of the corneal.[72] Several studies show that application of FGF2 facilitate epithelial wound closure.[73–75] Thus, the FGFs and FGFRs play important roles in many other ocular pathologies.

2. PDGFs in the Eye

The platelet-derived growth factor (PDGF) was first purifed in early 1970s from platelet extracts, and was biologically defined as a mitogen that can stimulate growth and migration of fibroblasts, smooth muscle cells and other types of mesenchymal cells.[76] Since then, tremendous progress has been achieved in understanding the complexity of the PDGF signaling network and the various roles PDGFs play during embryonic development, adult life, and in various pathologies. Furthermore, accumulating knowledge of physiological functions of PDGFs and their involvements in different diseases, such as retinal diseases, cardiovascular diseases, tumor development and growth, and fibrotic diseases, has shed new light onto the development of therapeutics targeting PDGF signaling pathways.

2.1. *The PDGF ligands and receptors*

The PDGFs are related to the VEGF family and form part of the cysteine-knot protein superfamily.[77] There are four members in the PDGF family, which are assembled as disulphide-bonded homodimers (PDGF-AA, -BB, -CC and -DD) or the PDGF-AB heterodimer.[78] PDGF-AA, -BB, and -AB are viewed as classical PDGFs, since they were isolated about three decades ago and have been intensively studied.[79] PDGF-CC and -DD were discovered nearly twenty years after the initial studies on PDGF-AA and -BB.[80,81] PDGF-CC and -DD differ from the classical PDGFs in that they are secreted as inactive precursor molecules due to the presence of an N-terminal CUB domain, which needs to be cleaved off by proteases to active them. Proteases can cleave the CUB domain, including plasmin, act on both PDGF-CC and -DD, while tissue plasminogen activator (tPA) and urokinase plasminogene activator (uPA) are specific for PDGF-CC and -DD, respectively.[77]

The four PDGFs share two receptors with tyrosine kinases (RTK) activities, namely PDGFR-α and PDGFR-β. The binding of a dimeric PDGF ligand triggers receptor dimerization in three forms, PDGFR-αα, -ββ and -αβ, which have different affinities for the four PDGF ligands[82]. PDGFR-αα binds to nearly all of the PDGF dimers, except PDGF-DD, while PDGFR-ββ only associates with PDGF-BB and PDGF-DD. As for PDGFR-αβ, it can be activated by almost all of the PDGFs except PDGF-AA. However, its binding affinity with PDGF-CC and -DD is markedly lower than with PDGF-AB and -BB. The binding of the PDGFs to the PDGFRs induces autophosphorylation of intracellular kinases domains, activating the downstream signaling cascades and propagating further signals via protein-protein interactions through specific domains, such as SH2, PTB, and PDZ[83]. The downstream signaling induced by the PDGFRs includes several well-known pathways, e.g., PI3K, Ras-MAPK, and PLC-γ, participating in numerous molecular and cellular aspects.[84]

2.2. *Expression and function of the PDGFs and PDGFRs in ocular cells in vitro*

In general, the PDGFs are expressed by many types of cells[79] and are major mitogens for cells of mesenchymal origin, such as smooth muscle cells and fibroblasts.[85] As early as in 1994, PDGF-AA, PDGF-BB and the two PDGF receptors have been shown to be present in cultured human retinal pigment epithelial cells (RPE).[86] In 2007, it was found that PDGF-CC and -DD mRNA levels are much higher than those of PDGF-AA and -BB in human RPE[87]. PDGF-CC, -DD, and -BB enhance RPE proliferation, while PDGF-DD, -BB, and -AB promote cell migration.[87] PDGF-BB also stimulates Müller cell proliferation through activating both c-JNK and PI3K/Akt pathways.[88] Both PDGFR-α and PDGFR-β are detected in rat retinal ganglion cells (RGCs), and their tyronsine phosphorylation mediated PI3K/Akt activation contributes to retinal neuronal cell survival.[89] In addition, PDGF-AA, -BB, and the two PDGF receptors are expressed in human corneal epithelial cells, stromal fibroblasts, and endothelial cells. PDGF-AA, -AB, and -BB promote the proliferation of corneal fibroblasts, and PDGF-BB has a chemotactic effect on them.[90]

2.3. *Normal distribution of the PDGFs and PDGFRs in the eye*

In the developing mouse retina, PDGF-AA and PDGFR-α transcripts are expressed in RGCs and astrocytes, respectively.[91] In addition, PDGF-AA is present in the corneal endothelium during embryonic development, and later on becomes restricted to the iris and ciliary body postnatally. PDGFR-α is present in the lens epithelium at all developmental stages.[92] In human, PDGF-AA and PDGFR-α expression is restricted to the retinal and choroidal vasculature.[93,94] PDGF-BB is predominantly expressed in the RGCs in mice and rat,[93,95] whereas in human retinae, it is detected in blood vessels and in the end-feet of Müller cells.[94] During mouse lens development, PDGF-BB is expressed in the iris and ciliary body, as well as in vascular cells surrounding the lens.[92] PDGF-CC is detected in the corneal epithelium and at the boundary of the eyelid in mouse embryos.[85] PDGF-CC is enriched in the RGC layer and the inner/outer nuclear layers in mouse retinae.[96] PDGF-DD is present in the iris and ciliary body, as well as in the limbal cells of the corneal epithelium. However, PDGF-DD is hardly found in the lens of neonatal rats eyes.[97] PDGF-DD is also localized in the adult rat retinae, where intense expression is found in the outer plexiform layer.[97] In human 6- and 7-week-old embryos, PDGFR-β is distributed in all the retinal layers, with the greatest signal in the lens epithelium, suggesting its participation in the development of human eyes.[98] Additionally, PDGFR-β is observed in the wall of the hyaloid blood vessels of the vitreous body in human embryos.[98] It is also found to be expressed in adult human retinal and choroid vessels[94] and in rat retina during development.[91]

2.4. *PDGFs and PDGFRs in eye development*

The PDGFs have been shown to be important for eye development. Lens- or RGC-specific expression of PDGF-A in mice promotes proliferation of astrocytes.[99,100] In addition, transgenic over-expression of PDGF-A induces lens epithelial cell proliferation, fiber cell differentiation, and cataracts in mice.[92] Over-expression of PDGF-A under the rhodopsin promoter in mice led to proliferation of glial cells and traction retinal detachment.[101] PDGF-BB blockade in rabbit corneal stroma decreases the generation of stromal αSMA$^+$ myofibroblasts. PDGF-BB therefore plays a role in the stromal-epithelial interaction and corneal stromal opacity.[102] Transgenic mice with genetic deletion of the PDGF-B retention motif not only exhibit pericyte loss and abnormal vasculature,[103] but also suffer from photoreceptor degeneration.[104] PDGF-B over-expression causes retinal abnormalities in mice. For instance, transgenic mice over-expressing PDGF-B under a nestin enhancer display retina defects, including disorganized nuclear layers, photoreceptor degeneration, and altered vasculature.[105] Similarly, folding of the retina and disorganization of the retinal capillaries also occur in MBP-specific PDGF-B expressing transgenic mice.[106] These data suggest that PDGF-B plays an essential role in retinal development. Several studies have also revealed the importance of PDGF-CC in retina maintenance. Both gain- and loss-of-function assays show that PDGF-CC has a protective effect on RGCs via PDGFR-α and PDGFR-β.[96] PDGF-CC treatment ameliorates photoreceptor degeneration without inducing retinal angiogenesis in mice.[107] In addition, mice with genetic deletion of PDGF-C exhibit exacerbated blood vessel degeneration mediated by PDGFR-α, PDGFR-β, and HMOX1.[108] As for PDGF-DD, its inhibition by neutralizing antibody significantly decreases the proliferation of rat lens epithelial cells in the anterior segments of the eye.[97] On the other hand, administration of PDGF-DD recombinant protein promotes the proliferation, survival, and migration of various ocular cells, such as retinal vascular pericytes, choroidal fibroblasts, and retinal-derived vascular endothelial cells,[109] suggesting an important role of PDGF-DD in the eye.

2.5. *Expression of the PDGFs and PDGFRs in eye diseases*

PDGF-CC is the predominant isoform observed in a rabbit model of proliferative vitreoretinopathy (PVR) and in the vitreous of PVR patients.[110] Furthermore, a greater percentage of PDGFR-α is activated than PDGFR-β in the membranes of PVR patients, suggesting important roles of PDGFR-α and PDGF-CC in PVR.[111] In retinal gliosis, the secretion of PDGF-AA and -BB is increased in the gliotic retinae compared with normal ones. This may contribute to the pathogenesis of fibrotic and degenerative ocular diseases.[112] In a traumatic tractional retinal detachment (TRD) rabbit model, PDGF-BB and -AB are present in Müller cells of the TRD eyes but not in normal eyes, indicating a role of PDGFs in TRD.[113]

2.6. *Roles of the PDGFs in ocular pathologies*

Diabetic retinopathy (DR) is a pathological condition that occurs as common blinding complication of diabetes. It affects retinal blood vessels and can lead to microanuerysms, hemorrhages, and edema.[114–116] Although DR is clinically apparent in vascular damages, it involves alteration in many other cell types of the retina as well, including ganglion cells, bipolar cells, amacrine cells, and pericytes.[117,118] Each type of these cells plays multiple roles in contributing to the development of DR. One of the key features of DR is characterized by pericyte loss, which consequently induces apoptosis of retinal pericytes.[114,119] In normal retinal capillaries, the ratio of pericytes to endothelial cells is about one. In DR patients, however, the ratio is less than one because of pericyte loss.[120] In addition, the number of pericytes in diabetic WT mice decreases by about 40% compared with that of non-diabetic WT mice.

The loss of pericyte is pronounced in streptozotocin (STZ)-induced diabetic PDGF-$B^{+/-}$ mice with a 3.5-fold increase of acellular capillaries, suggesting that pericyte coverage is critical for the survival of endothelial cells in diabetes.[121] In a STZ-induced diabetic rat model, PDGF-B mRNA level is upregulated by protein kinase C (PKC).[122] Moreover, it has been shown that both PKCδ and p38α-MAPK are activated in hyperglycemia, which upregulates SHP-1 phosphatase, and then PDGFR-β dephosphorylation. These subsequently increase pericyte apoptosis in cultured cells *in vitro* and in animal models of diabetes *in vivo*.[123] PDGF-AB was found to be significantly increased in the vitreous of patients with proliferative diabetic retinopathy (PDR). The vitreous levels of the PDGFs

(PDGF-AA, PDGF-BB, and PDGF-AB) correlate differentially with PDR.[124,125] and PDGF-BB has been identified as a therapeutic target for PDR in a systemic meta-analysis. Moreover, it has been shown that combined inhibition of VEGF-A and PDGF-BB reduced PDR better than single inhibition of VEGF-A or PDGF-BB alone,[126] supporting a role of PDGF-BB in PDR.

Proliferative vitreoretinopathy (PVR) is an ocular disease that occurs in the setting of a rhegmatogenous retinal detachment. The development of PVR is most often associated with formation of retinal holes or tears.[127,128] In PVR, after retinal detachment, cells depart their normal location in the retina and migrate to the retinal surface. Retinal pigmented epithelial (RPE) cells have been considered a key player in the pathogenesis of PVR. Under normal conditions, RPE cells do not have direct physical contact with the vitreous. However, in PVR, RPE cells can be found in nearly all PVR epiretinal membranes.[129] Both PDGFs and PDGFRs are expressed in the epiretinal membranes isolated from PVR patients.[87,94] In addition to RPE cells, it has been recently appreciated that Müller glia also play a role in the pathogenesis of PVR.[112] A number of cytokines and inflammatory factors, such as G-CSF, MCP-1, PDGF-BB, RANTES, VEGF, and TGFβ2, are produced by Müller glia, suggesting that Müller glia is a critical source of cytokines and growth factors associated with retinal gliosis in PVR. The two PDGFRs play important roles in the pathological processes of PVR. It has been shown that in a rabbit model, PVR can be hardly induced by vitreous injection of mouse embryo fibroblasts (MEF) with genetic deletion of PDGFR-α or PDGFR-β.[130] Other studies have demonstrated that PDGFR-α is more abundantly expressed in PVR than PDGFR-β.[131] Moreover, it has been suggested that PVR is induced by PI3K and PLCγ signaling pathway via PDGFR-α.[131]

Transgenic mice expressing PDGF-A or PDGF-B under a rhodopsin promoter display different patterns of retinopathies. Overexpression of PDGF-A in the retina results in extensive proliferation of glial cells and retinal detachment without engagement of vascular cells, which is similar to human PVR. However, overexpression of PDGF-B leads to traction of vascular and non-vascular cells, which resembles DR in humans.[101] Furthermore, it was found that PDGF-C is the predominant PDGF among others detected in PVR models *in vitro* and *in vivo*.[110,132] Likewise, PDGF-C is detected in PVR patients with a concentration of 50–1000 ng/ml.[110] Activation of PDGFR-α suppresses p53 activity, which promotes cell survive in response to stress such as hypoxia.[133] Subsequent studies have suggested that VEGF-A in the vitreous can block PDGF-dependent binding and activation of PDGFRs, thus affecting its signaling pathways and cellular responses.[134] A recent study has shown that PDGFR-α can be activated by the NADPH oxidase-Src pathway, which in turn leads to PI3K/Akt/mTORC1 signaling to inhibit autophagy.[135]

Age-related macular degeneration (AMD) is a disease associated with a deterioration of central vision. It is an unfortunately common and predominantly affects the elderly population. AMD can be classified as non-neovascular (dry or non-exudative) AMD or neovascular (wet or exudative) AMD.[136] The complication of neovascular AMD is driven by choroidal neovascularization (CNV), in which new blood vessels grow from the choroid (choriocapillaries) through the Bruch's membrane under the RPE layer.[137,138] Currently, anti-VEGF therapies can halt vision loss in some patients with AMD. However, not all AMD patients can benefit, implicating that VEGF-independent pathways exist. Indeed, blocking the PDGF-B signaling pathway using an anti-PDGFR-β antibody enhanced the effect of anti-VEGF therapy in CNV.[139] Indeed, it has been shown that blocking both VEGF-A and PDGF-B simultaneously is more effective in inhibiting CNV.[140]

It has also been reported that inhibition of PDGF-C suppressed both choroidal and retinal neovascularization.[141] Additionally, PDGF-C targeting affected different types of cells important for pathological angiogenesis, including endothelial cells, vascular smooth muscle cells, choroidal fibroblasts, RPEs, and macrophages. Mechanistically, glycogen synthase kinase-3β (GSK-3β) phosphorylation is shown to be regulated by PDGF-C. It will be interesting to study whether combined inhibition of VEGF-A and PDGF-C would display a better effect to treat neovascular AMD compared with monotherapy.

References

1. Chen, G.J. and Forough, R. Fibroblast growth factors, fibroblast growth factor receptors, diseases, and drugs. *Recent Pat Cardiovasc Drug Discov* **1**, 211–224 (2006).
2. Noji, S., Matsuo, T., Koyama, E., *et al.* Expression pattern of acidic and basic fibroblast growth factor genes in adult rat eyes. *Biochem Biophys Res Commun* **168**, 343–349 (1990).
3. Tcheng, M., Fuhrmann, G., Hartmann, MP., *et al.* Spatial and temporal expression patterns of FGF receptor genes type 1 and type 2 in the developing chick retina. *Expe Eye Res* **58**, 351–358 (1994).
4. Fuhrmann, V., Kinkl, N., Leveillard, T., *et al.* Fibroblast growth factor receptor 4 (FGFR4) is expressed in adult rat and human retinal photoreceptors and neurons. *J Mol Neurosci* **13**, 187–197 (1999).
5. Lovicu, F.J. and McAvoy, J.W. Localization of acidic fibroblast growth factor, basic fibroblast growth factor, and heparan sulphate proteoglycan in rat lens: Implications for lens polarity and growth patterns. *Invest Ophthalmol Vis Sci* **34**, 3355–3365 (1993).
6. McAvoy, J.W. and Chamberlain, C.G. Fibroblast growth factor (FGF) induces different responses in lens epithelial cells depending on its concentration. *Development* **107**, 221–228 (1989).
7. de Iongh, R. and McAvoy, J.W. Spatio-temporal distribution of acidic and basic FGF indicates a role for FGF in rat lens morphogenesis. *Developmental Dynamics: An Official Publication of the American Association of Anatomists* **198**, 190–202 (1993).
8. Orr-Urtreger, A., Givol, D., Yayon, A., Yarden, Y., *et al.* Developmental expression of two murine fibroblast growth factor receptors, flg and bek. *Development* **113**, 1419–1434 (1991).
9. Launay, C., Fromentoux, V., Thery, C., *et al.* Comparative analysis of the tissue distribution of three fibroblast growth factor receptor mRNAs during amphibian morphogenesis. *Differentiation: Research in Biological Diversity* **58**, 101–111 (1994).
10. Valter, K., van Driel, D., Bisti, S., and Stone, J. FGFR1 expression and FGFR1-FGF-2 colocalisation in rat retina: Sites of FGF-2 action on rat photoreceptors. *Growth Factors* **20**, 177–188 (2002).
11. Malecaze, F., Mathis, A, Arne. J.L., Raulais. D., *et al.* Localization of acidic fibroblast growth factor in proliferative vitreoretinopathy membranes. *Curr Eye Res* **10**, 719–729 (1991).

12. Fredj-Reygrobellet, D., Baudouin, C., Negre, F., *et al.* Acidic FGF and other growth factors in preretinal membranes from patients with diabetic retinopathy and proliferative vitreoretinopathy. *Ophthalmic Res* **23**, 154–161 (1991).
13. Wen, R., Song, Y., Cheng, T., *et al.* Injury-induced upregulation of bFGF and CNTF mRNAS in the rat retina. *J Neurosci* **15**, 7377–7385 (1995).
14. Ozaki, S., Radeke, M.J., Anderson, D.H. Rapid upregulation of fibroblast growth factor receptor 1 (flg) by rat photoreceptor cells after injury. *Invest Ophthalmol Vis Sci* **41**, 568–579 (2000).
15. Gao, H., Hollyfield, JG. Basic fibroblast growth factor in retinal development: Differential levels of bFGF expression and content in normal and retinal degeneration (rd) mutant mice. *Dev Biol* **169**, 168–184 (1995).
16. Yamamoto, C., Ogata, N., Matsushima, M., *et al.* Gene expressions of basic fibroblast growth factor and its receptor in healing of rat retina after laser photocoagulation. *Jpn J Ophthalmol* **40**, 480–490 (1996).
17. Valter, K., Maslim, J., Bowers, F., Stone, J. Photoreceptor dystrophy in the RCS rat: roles of oxygen, debris, and bFGF. *Invest Ophthalmol Vis Sci* **39**, 2427–2442 (1998).
18. Fang, L., Barber, A.J., Shenberger, J.S. Regulation of fibroblast growth factor 2 expression in oxygen-induced retinopathy. *Invest Ophthalmol Vis Sci* **56**, 207–215 (2015).
19. Dorrell, M., Uusitalo-Jarvinen, H., Aguilar, E., Friedlander, M. Ocular neovascularization: Basic mechanisms and therapeutic advances. *Surv Ophthalmol* **52** (**Suppl 1**), S3-S19 (2007).
20. Hartnett, M.E. and Penn, J.S. Mechanisms and management of retinopathy of prematurity. *N Engl J Med* **367**, 2515–2526 (2012).
21. Afzal, A., Shaw, L.C., Ljubimov, A.V., Boulton, M.E., Segal, M.S., Grant, M.B. Retinal and choroidal microangiopathies: therapeutic opportunities. *Microvasc Res* **74**, 131–144 (2007).
22. Klein, R., Wang, Q., Klein, B.E., Moss, S.E., *et al.* The relationship of age-related maculopathy, cataract, and glaucoma to visual acuity. *Invest Ophthalmol Vis Sci* **36**, 182–191 (1995).
23. Coats, D.K. Retinopathy of prematurity: involution, factors predisposing to retinal detachment, and expected utility of preemptive surgical reintervention. *Trans Am Ophthalmol Soc* **103**, 281–312 (2005).
24. Sapieha, P., Joyal, J.S., Rivera, J.C., *et al.* Retinopathy of prematurity: understanding ischemic retinal vasculopathies at an extreme of life. *J Clin Invest* **120**, 3022–3032 (2010).

25. Markowska, A.I., Cao, Z., Panjwani, N. Glycobiology of ocular angiogenesis. *Glycobiology* **24**, 1275–1282 (2014).
26. van Lookeren Campagne, M., LeCouter, J., Yaspan, B.L., Ye, W. Mechanisms of age-related macular degeneration and therapeutic opportunities. *J Pathol* **232**, 151–164 (2014).
27. Shakiba, Y., Mansouri, K., Arshadi, D., Rezaei, N. Corneal neovascularization: molecular events and therapeutic options. *Recent Pat Inflamm Allergy Drug Discov* **3**, 221–231 (2009).
28. Presta, M., Dell'Era, P., Mitola, S., *et al.* Fibroblast growth factor/fibroblast growth factor receptor system in angiogenesis. *Cytokine Growth Factor Rev* **16**, 159–178 (2005).
29. Moscatelli, D., Presta, M., Rifkin, D.B. Purification of a factor from human placenta that stimulates capillary endothelial cell protease production, DNA synthesis, and migration. *Proc Natl Acad Sci USA* **83**, 2091–2095 (1986).
30. Klein, S., Giancotti, F.G., Presta, M., Albelda, S.M., *et al.* Basic fibroblast growth factor modulates integrin expression in microvascular endothelial cells. *Mol Bio Cell* **4**:973–982 (1993).
31. Stavri, G.T., Zachary, I.C., Baskerville, P.A., Martin, J.F., *et al.* Basic fibroblast growth factor upregulates the expression of vascular endothelial growth factor in vascular smooth muscle cells. Synergistic interaction with hypoxia. *Circulation* **92**:11–14 (1995).
32. Seghezzi, G., Patel, S., Ren, C.J., *et al.* Fibroblast growth factor-2 (FGF-2) induces vascular endothelial growth factor (VEGF) expression in the endothelial cells of forming capillaries: An autocrine mechanism contributing to angiogenesis. *J Cell Biol* **141**:1659–1673 (1998).
33. Stahl, A., Paschek, L., Martin, G., Feltgen, N., *et al.* Combinatory inhibition of VEGF and FGF2 is superior to solitary VEGF inhibition in an *in vitro* model of RPE-induced angiogenesis. *Albrecht Von Graefes Arch Klin Exp Ophthalmol* **247**:767–773 (2009).
34. Tobe, T., Ortega, S., Luna, J.D., *et al.* Targeted disruption of the FGF2 gene does not prevent choroidal neovascularization in a murine model. *Am J Pathol* **153**:1641–1646 (1998).
35. Amin, R., Puklin, J.E., Frank, R.N. Growth factor localization in choroidal neovascular membranes of age-related macular degeneration. *Invest Ophthalmol Vis Sci* **35**:3178–3188 (1994).
36. Wong, C.G., Rich, K.A., Liaw, L.H., Hsu, H.T., Berns, M.W. Intravitreal VEGF and bFGF produce florid retinal neovascularization and hemorrhage in the rabbit. *Curr Eye Res* **22**:140–147 (2001).

37. Sivalingam, A., Kenney, J., Brown, G.C., Benson, W.E., Donoso, L. Basic fibroblast growth factor levels in the vitreous of patients with proliferative diabetic retinopathy. *Arch Ophthal* **108**:869–872 (1990).
38. Mitamura, Y., Harada, C., Harada, T. Role of cytokines and trophic factors in the pathogenesis of diabetic retinopathy. *Curr Diabetes Rev* **1**:73–81 (2005).
39. Zakareia, F.A., Alderees, A.A., Al Regaiy, K.A., Alrouq, F.A. Correlation of electroretinography b-wave absolute latency, plasma levels of human basic fibroblast growth factor, vascular endothelial growth factor, soluble fatty acid synthase, and adrenomedullin in diabetic retinopathy. *J Diabetes Complications* **24**:179–185 (2010).
40. Hueber, A., Wiedemann, P., Esser, P., Heimann, K. Basic fibroblast growth factor mRNA, bFGF peptide and FGF receptor in epiretinal membranes of intraocular proliferative disorders (PVR and PDR). *Int Ophthalmol* **20**, 345–350 (1996).
41. Harada, T., Harada, C., Kohsaka, S., *et al.* Microglia-Muller glia cell interactions control neurotrophic factor production during light-induced retinal degeneration. *J Neurosci* **22**, 9228–9236 (2002).
42. Schneeberger, S.A., Hjelmeland, L.M., Tucker, R.P., Morse, L.S. Vascular endothelial growth factor and fibroblast growth factor 5 are colocalized in vascular and avascular epiretinal membranes. *Ame J Ophthalmol* **124**, 447–454 (1997).
43. Hajrasouliha, A.R., Sadrai, Z., Chauhan, S.K., Dana, R. b-FGF induces corneal blood and lymphatic vessel growth in a spatially distinct pattern. *Cornea* **31**, 804–809 (2012).
44. Cao, R., Eriksson, A., Kubo, H., Alitalo, K., *et al.* Comparative evaluation of FGF-2-, VEGF-A-, and VEGF-C-induced angiogenesis, lymphangiogenesis, vascular fenestrations, and permeability. *Circ Res* **94**, 664–670 (2004).
45. Gillis, P., Savla, U., Volpert, O.V., *et al.* Keratinocyte growth factor induces angiogenesis and protects endothelial barrier function. *J Cell Sci* **112 (Pt 12)**, 2049–2057 (1999).
46. Zhou, M., Sutliff, R.L., Paul, R.J., *et al.* Fibroblast growth factor 2 control of vascular tone. *Nat Med* **4**, 201–207 (1998).
47. Ozaki, H., Okamoto, N., Ortega, S., *et al.* Basic fibroblast growth factor is neither necessary nor sufficient for the development of retinal neovascularization. *Am J Pathol* **153**, 757–765 (1998).
48. Miller, DL., Ortega, S., Bashayan, O., Basch, R., *et al.* Compensation by fibroblast growth factor 1 (FGF1) does not account for the mild

phenotypic defects observed in FGF2 null mice. *Mol Cell Bio* **20**, 2260–2268 (2000).
49. Deng, CX., Wynshaw-Boris, A., Shen, M.M., Daugherty, C., *et al.* Murine FGFR-1 is required for early postimplantation growth and axial organization. *Genes Dev* **8**, 3045–3057 (1994).
50. Yamaguchi, T.P., Harpal, K., Henkemeyer, M., Rossant, J. FGFR-1 is required for embryonic growth and mesodermal patterning during mouse gastrulation. *Genes Dev* **8**, 3032–3044 (1994).
51. Xu, X., Weinstein, M., Li, C., *et al.* Fibroblast growth factor receptor 2 (FGFR2)-mediated reciprocal regulation loop between FGF8 and FGF10 is essential for limb induction. *Development* **125**, 753–765 (1998).
52. Weinstein, M., Xu, X., Ohyama, K., Deng, C.X. FGFR-3 and FGFR-4 function cooperatively to direct alveogenesis in the murine lung. *Development* **125**, 3615–3623 (1998).
53. Colvin. J.S., Bohne, B.A., Harding, G.W., *et al.* Skeletal overgrowth and deafness in mice lacking fibroblast growth factor receptor 3. *Nat Genet* **12**, 390–397 (1996).
54. Weikel, KA., Garber, C., Baburins, A., and Taylor, A. Nutritional modulation of cataract. *Nutr Rev* **72**, 30–47 (2014).
55. Brian, G. and Taylor, H. Cataract blindness — challenges for the 21st century. *Bull World Health Organ* **79**, 249–256 (2001).
56. Awasthi, N., Guo, S., and Wagner, B.J. Posterior capsular opacification: a problem reduced but not yet eradicated. *Arch Ophthalmol* **127**, 555–562 (2009).
57. Spalton, D. Posterior capsule opacification: Have we made a difference? *Br J Ophthalmol* **97**, 1–2 (2013).
58. Robinson, M.L. An essential role for FGF receptor signaling in lens development. *Sem Cell Dev Biol* **17**, 726–740 (2006).
59. Lovicu, F.J., Chamberlain, C.G., and McAvoy, J.W. Differential effects of aqueous and vitreous on fiber differentiation and extracellular matrix accumulation in lens epithelial explants. *Invest Ophthalmol Vis Sci* **36**, 1459–1469 (1995).
60. Lovicu, F.J. and Overbeek, P.A. Overlapping effects of different members of the FGF family on lens fiber differentiation in transgenic mice. *Development* **125**, 3365–3377 (1998).
61. Robinson, M.L., Overbeek, P.A., Verran, D.J., *et al.* Extracellular FGF-1 acts as a lens differentiation factor in transgenic mice. *Development* **121**, 505–514 (1995).

62. Coulombre, J.L. and Coulombre, A.J. Lens Development: Fiber Elongation and Lens Orientation. *Science* **142**, 1489–1490 (1963).
63. Zhao, H., Yang, T., Madakashira, B.P., *et al.* Fibroblast growth factor receptor signaling is essential for lens fiber cell differentiation. *Dev Bio* **318**, 276–288 (2008).
64. Iyengar, L., Patkunanathan, B., McAvoy, J.W., and Lovicu, F.J. Growth factors involved in aqueous humour-induced lens cell proliferation. *Growth Factors* **27**, 50–62 (2009).
65. Mansfield, K.J., Cerra, A., Chamberlain, C.G. FGF-2 counteracts loss of TGFbeta affected cells from rat lens explants: implications for PCO (after cataract). *Mol Vis* **10**, 521–532 (2004).
66. Wormstone, I.M., Del Rio-Tsonis, K., McMahon, G., *et al.* FGF: An autocrine regulator of human lens cell growth independent of added stimuli. *Invest Ophthalmol Vis Sci* **42**, 1305–1311 (2001).
67. Faktorovich, E.G., Steinberg, R.H., Yasumura, D., *et al.* Photoreceptor degeneration in inherited retinal dystrophy delayed by basic fibroblast growth factor. *Nature* **347**, 83–86 (1990).
68. Faktorovich, E.G., Steinberg, R.H., Yasumura, D., *et al.* Basic fibroblast growth factor and local injury protect photoreceptors from light damage in the rat. *J Neurosci* **12**, 3554–3567 (1992).
69. Lau, D., McGee, L.H., Zhou, S., *et al.* Retinal degeneration is slowed in transgenic rats by AAV-mediated delivery of FGF-2. *Invest Ophthalmol Vis Sci* **41**, 3622–3633 (2000).
70. Green, E.S., Rendahl, K.G., Zhou, S., *et al.* Two animal models of retinal degeneration are rescued by recombinant adeno-associated virus-mediated production of FGF-5 and FGF-18. *Mol Therapy* **3**, 507–515 (2001).
71. Siffroi-Fernandez, S., Felder-Schmittbuhl, M.P., Khanna, H., *et al.* FGF19 exhibits neuroprotective effects on adult mammalian photoreceptors *in vitro*. *Invest Ophthalmol Vis Sci* **49**, 1696–1704 (2008).
72. Baldwin, H.C. and Marshall, J. Growth factors in corneal wound healing following refractive surgery: A review. *Acta Ophthalmol Scand* **80**, 238–247 (2002).
73. Hu, C., Ding, Y., Chen, J., *et al.* Basic fibroblast growth factor stimulates epithelial cell growth and epithelial wound healing in canine corneas. *Vet Ophthalmol* **12**, 170–175 (2009).
74. Grant, M.B., Khaw, P.T., Schultz, G.S., *et al.* Effects of epidermal growth factor, fibroblast growth factor, and transforming growth factor-beta on corneal cell chemotaxis. *Invest Ophthalmol Vis Sci* **33**, 3292–3301 (1992).

75. Pancholi, S., Tullo, A., Khaliq, A., *et al.*. The effects of growth factors and conditioned media on the proliferation of human corneal epithelial cells and keratocytes. *Albrecht Von Graefes Arch Klin Exp Ophthalmol* **236**, 1–8 (1998).
76. Ross, R., Glomset, J., Kariya, B., Harker, L. A platelet-dependent serum factor that stimulates the proliferation of arterial smooth muscle cells *in vitro*. *Proc Natl Acad Sci USA* **71**, 1207–1210 (1974).
77. Fredriksson, L., Li, H., Eriksson, U. The PDGF family: Four gene products form five dimeric isoforms. *Cytokine Growth Fact Rev* **15**, 197–204 (2004).
78. Boor, P., Ostendorf, T., Floege, J. PDGF and the progression of renal disease. *Nephrol Dial Transpl* **29**(**Suppl 1**), i45–i54 (2014).
79. Heldin, C.H. and Westermark, B. Mechanism of action and in vivo role of platelet-derived growth factor. *Physiol Rev* **79**, 1283–1316 (1999).
80. Li, X., Ponten, A., Aase, K., *et al.* PDGF-C is a new protease-activated ligand for the PDGF alpha-receptor. *Nat Cell Biol* **2**, 302–309 (2000).
81. Bergsten, E., Uutela, M., Li, X., *et al.* PDGF-D is a specific, protease-activated ligand for the PDGF beta-receptor. *Nat Cell Biol* **3**, 512–516 (2001).
82. Demoulin, J.B., Essaghir, A. PDGF receptor signaling networks in normal and cancer cells. *Cytokine Growth Fact Rev* **25**, 273–283 (2014).
83. Andrae, J., Gallini, R., Betsholtz, C. Role of platelet-derived growth factors in physiology and medicine. *Genes Dev* **22**, 1276–1312 (2008).
84. Tallquist, M., Kazlauskas, A. PDGF signaling in cells and mice. *Cytokine Growth Fact Rev* **15**, 205–213 (2004).
85. Aase, K., Abramsson, A., Karlsson, L., *et al.* Expression analysis of PDGF-C in adult and developing mouse tissues. *Mech Dev* **110**, 187–191 (2002).
86. Campochiaro, P.A., Hackett, S.F., Vinores, S.A., *et al.* Platelet-derived growth factor is an autocrine growth stimulator in retinal pigmented epithelial cells. *J Cell Sci* **107 (Pt 9)**, 2459–2469 (1994).
87. Cui, J.Z., Chiu, A., Maberley, D., *et al.* Stage specificity of novel growth factor expression during development of proliferative vitreoretinopathy. *Eye (Lond)* **21**, 200–208 (2007).
88. Moon, S.W., Chung, E.J., Jung, S.A., Lee, J.H. PDGF stimulation of Muller cell proliferation: Contributions of c-JNK and the PI3K/Akt pathway. *Biochem Biophys Res Commun* **388**, 167–171 (2009).
89. Biswas, S.K., Zhao, Y., Nagalingam, A., Gardner, T.W., Sandirasegarane, L. PDGF- and insulin/IGF-1-specific distinct modes of class IA PI

3-kinase activation in normal rat retinas and RGC-5 retinal ganglion cells. *Invest Ophthalmol Vis Sci* **49**, 3687–3698 (2008).

90. Kim, W.J., Mohan, R.R., Mohan, R.R., Wilson, S.E. Effect of PDGF, IL-1alpha, and BMP2/4 on corneal fibroblast chemotaxis: Expression of the platelet-derived growth factor system in the cornea. *Invest Ophthalmol Vis Sci* **40**, 1364–1372 (1999).
91. Mudhar, H.S., Pollock, R.A., Wang, C., *et al.* PDGF and its receptors in the developing rodent retina and optic nerve. *Development* **118**, 539–552 (1993).
92. Reneker, L.W. and Overbeek, P.A. Lens-specific expression of PDGF-A alters lens growth and development. *Dev Biol* **180**, 554–565 (1996).
93. Cox, O.T., Simpson, D.A., Stitt, A.W., Gardiner, T.A. Sources of PDGF expression in murine retina and the effect of short-term diabetes. *Mol Vis* **9**, 665–672 (2003).
94. Robbins, S.G., Mixon, R.N., Wilson, D.J., *et al.* Platelet-derived growth factor ligands and receptors immunolocalized in proliferative retinal diseases. *Invest Ophthalmol Vis Sci* **35**, 3649–3663 (1994).
95. Mekada, A., Sasahara, M., Yamada, E., Kani, K., Hazama, F. Platelet-derived growth factor B-chain expression in the rat retina and optic nerve: Distribution and changes after transection of the optic nerve. *Vision Res* **38**, 3031–3039 (1998).
96. Tang, Z., Arjunan, P., Lee, C., *et al.* Survival effect of PDGF-CC rescues neurons from apoptosis in both brain and retina by regulating GSK3beta phosphorylation. *J Exp Med* **207**, 867–880 (2010).
97. Ray, S., Gao, C., Wyatt, K., *et al.* Platelet-derived growth factor D, tissue-specific expression in the eye, and a key role in control of lens epithelial cell proliferation. *J Biol Chem* **280**, 8494–8502 (2005).
98. Bozanic, D., Bocina, I., Saraga-Babic, M. Involvement of cytoskeletal proteins and growth factor receptors during development of the human eye. *Anat Embryol (Berl)* **211**, 367–377 (2006).
99. Reneker, L.W. and Overbeek, P.A. Lens-specific expression of PDGF-A in transgenic mice results in retinal astrocytic hamartomas. *Invest Ophthalmol Vis Sci* **37**, 2455–2466 (1996).
100. Fruttiger, M., Calver, A.R., Kruger, W.H., *et al.* PDGF mediates a neuron-astrocyte interaction in the developing retina. *Neuron* **17**, 1117–1131 (1996).
101. Mori, K., Gehlbach, P., Ando, A., *et al.* Retina-specific expression of PDGF-B versus PDGF-A: vascular versus nonvascular proliferative retinopathy. *Invest Ophthalmol Vis Sci* **43**, 2001–2006 (2002).

102. Kaur, H., Chaurasia, S.S., de Medeiros, F.W., *et al.* Corneal stroma PDGF blockade and myofibroblast development. *Exp Eye Res* **88**, 960–965 (2009).
103. Lindblom, P., Gerhardt, H., Liebner, S., *et al.* Endothelial PDGF-B retention is required for proper investment of pericytes in the microvessel wall. *Genes Dev* **17**, 1835–1840 (2003).
104. Genove, G., Mollick, T., Johansson, K. Photoreceptor degeneration, structural remodeling and glial activation: a morphological study on a genetic mouse model for pericyte deficiency. *Neuroscience* **279**, 269–284 (2014).
105. Edqvist, P.H., Niklasson, M., Vidal-Sanz, M., *et al.* Platelet-derived growth factor over-expression in retinal progenitors results in abnormal retinal vessel formation. *PLoS One* **7**, e42488 (2012).
106. Forsberg-Nilsson, K., Erlandsson, A., Zhang, X.Q., *et al.* Oligodendrocyte precursor hypercellularity and abnormal retina development in mice overexpressing PDGF-B in myelinating tracts. *Glia* **41**, 276–289 (2003).
107. Wang, Y., Abu-Asab, M.S., Yu, C.R., *et al.* Platelet-derived growth factor (PDGF)-C inhibits neuroretinal apoptosis in a murine model of focal retinal degeneration. *Lab Invest* **94**, 674–682 (2014).
108. He, C., Zhao, C., Kumar, A., *et al.* Vasoprotective effect of PDGF-CC mediated by HMOX1 rescues retinal degeneration. *Proc Natl Acad Sci USA* **111**, 14806–14811 (2014).
109. Kumar, A., Hou, X., Lee, C., *et al.* Platelet-derived growth factor-DD targeting arrests pathological angiogenesis by modulating glycogen synthase kinase-3beta phosphorylation. *J Biol Chem* **285**, 15500–15510 (2010).
110. Lei, H., Hovland, P., Velez, G., *et al.* A potential role for PDGF-C in experimental and clinical proliferative vitreoretinopathy. *Invest Ophthalmol Vis Sci* **48**, 2335–2342 (2007).
111. Cui, J., Lei, H., Samad, A., *et al.* PDGF receptors are activated in human epiretinal membranes. *Exp Eye Res* **88**:438–444 (2009).
112. Eastlake, K., Banerjee, P.J., Angbohang, A., *et al.* Muller glia as an important source of cytokines and inflammatory factors present in the gliotic retina during proliferative vitreoretinopathy. *Glia* 2015.
113. Westra, I., Robbins, S.G., Wilson, D.J., *et al.* Time course of growth factor staining in a rabbit model of traumatic tractional retinal detachment. *Graefes Arch Clin Exp Ophthalmol* **233**, 573–581 (1995).

114. Shin, E.S., Sorenson, C.M., Sheibani, N. Diabetes and retinal vascular dysfunction. *J Ophthalmic Vis Res* **9**, 362–373 (2014).
115. Nentwich, M.M. and Ulbig, M.W. Diabetic retinopathy — ocular complications of diabetes mellitus. *World J Diabetes* **6**, 489–499 (2015).
116. Semeraro, F., Cancarini, A., dell'Omo, R., Rezzola, S., Romano, M.R., Costagliola, C. Diabetic Retinopathy: Vascular and Inflammatory Disease. *J Diabetes Res* **2015**, 582060 (2015).
117. Kern, T..S. and Barber, A..J. Retinal ganglion cells in diabetes. *J Physiol* **586**, 4401–4408 (2008).
118. Vujosevic, S. and Midena, E. Retinal layers changes in human preclinical and early clinical diabetic retinopathy support early retinal neuronal and Muller cells alterations. *J Diabetes Res* **2013**, 905058 (2013).
119. Armulik, A., Abramsson, A., Betsholtz, C. Endothelial/pericyte interactions. *Circ Res* **97**, 512–523 (2005).
120. Armulik, A., Genove, G., Betsholtz, C. Pericytes: Developmental, physiological, and pathological perspectives, problems, and promises. *Dev Cell* **21**, 193–215 (2011).
121. Hammes, H P., Lin, J., Renner, O., *et al.* Pericytes and the pathogenesis of diabetic retinopathy. *Diabetes* **51**, 3107–3112 (2002).
122. Yokota, T., Ma, R. C., Park, J. Y., *et al.* Role of protein kinase C on the expression of platelet-derived growth factor and endothelin-1 in the retina of diabetic rats and cultured retinal capillary pericytes. *Diabetes* **52**, 838–845 (2003).
123. Geraldes, P., Hiraoka-Yamamoto, J., Matsumoto, M., *et al.* Activation of PKC-delta and SHP-1 by hyperglycemia causes vascular cell apoptosis and diabetic retinopathy. *Nat Med* **15**, 1298–1306 (2009).
124. Praidou, A., Klangas, I., Papakonstantinou, E., *et al.* Vitreous and serum levels of platelet-derived growth factor and their correlation in patients with proliferative diabetic retinopathy. *Curr Eye Res* **34**, 152–161 (2009).
125. Praidou, A., Papakonstantinou, E., Androudi, S., *et al.* Vitreous and serum levels of vascular endothelial growth factor and platelet-derived growth factor and their correlation in patients with non-proliferative diabetic retinopathy and clinically significant macula oedema. *Acta Ophthalmol* **89**, 248–254 (2011).
126. McAuley, A. K., Sanfilippo, P. G., Hewitt, A. W., *et al.* Vitreous biomarkers in diabetic retinopathy: a systematic review and meta-analysis. *J Diabetes Complications* **28**, 419–425 (2014).

127. Lei, H., Rheaume, M. A., Kazlauskas, A. Recent developments in our understanding of how platelet-derived growth factor (PDGF) and its receptors contribute to proliferative vitreoretinopathy. *Exp Eye Res* **90**, 376–381 (2010).
128. Pennock, S., Haddock, L. J., Eliott, D., Mukai, S., Kazlauskas, A. Is neutralizing vitreal growth factors a viable strategy to prevent proliferative vitreoretinopathy? *Prog Retin Eye Res* **40**, 16–34 (2014).
129. Leaver, P.K. Proliferative vitreoretinopathy. *Br J Ophthalmol* **79**, 871–872 (1995).
130. Andrews, A., Balciunaite, E., Leong, F.L., *et al.* Platelet-derived growth factor plays a key role in proliferative vitreoretinopathy. *Invest Ophthalmol Vis Sci* **40**, 2683–2689 (1999).
131. Ikuno, Y., Leong, F.L., Kazlauskas, A. PI3K and PLCgamma play a central role in experimental PVR. *Invest Ophthalmol Vis Sci* **43**, 483–489 (2002).
132. Lei, H., Velez, G., Hovland, P., Hirose, T., Kazlauskas, A. Plasmin is the major protease responsible for processing PDGF-C in the vitreous of patients with proliferative vitreoretinopathy. *Invest Ophthalmol Vis Sci* **49**, 42–48 (2008).
133. Lei, H., Velez, G., Kazlauskas, A. Pathological signaling via platelet-derived growth factor receptor {alpha} involves chronic activation of Akt and suppression of p53. *Mol Cell Bio* **31**, 1788–1799 (2011).
134. Lei, H., Rheaume, M.A., Cui, J., *et al.* A novel function of p53: a gatekeeper of retinal detachment. *Am J Pathol* **181**, 866–874 (2012).
135. Lei, H. and Kazlauskas, A. A reactive oxygen species-mediated, self-perpetuating loop persistently activates platelet-derived growth factor receptor alpha. *Mol Cell Biol* **34**, 110–122 (2014).
136. Green, W.R. and Enger, C. Age-related macular degeneration histopathologic studies. The 1992 Lorenz E. Zimmerman Lecture. *Ophthalmology* **100**, 1519–1535 (1993).
137. Campochiaro, P.A. Retinal and choroidal neovascularization. *J Cell Physio* **184**, 301–310 (2000).
138. Campochiaro, P.A. Molecular pathogenesis of retinal and choroidal vascular diseases. *Pro Retin Eye Res* **49**, 67–81 (2015).
139. Jo, N., Mailhos, C., Ju, M., *et al.* Inhibition of platelet-derived growth factor B signaling enhances the efficacy of anti-vascular endothelial growth factor therapy in multiple models of ocular neovascularization. *Am J Pathol* **168**, 2036–2053 (2006).

140. Dong, A., Seidel, C., Snell, D., *et al.* Antagonism of PDGF-BB suppresses subretinal neovascularization and enhances the effects of blocking VEGF-A. *Angiogenesis* **17**, 553–562 (2014).
141. Hou, X., Kumar, A., Lee, C., *et al.* PDGF-CC blockade inhibits pathological angiogenesis by acting on multiple cellular and molecular targets. *Proc Natl Acad Sci USA* **107**, 12216–12221 (2010).

Chapter 10

FGF Ligand Traps for the Therapy of FGF-Dependent Tumors

Marco Rusnati*, Marco Presta* and Roberto Ronca*,†

Abstract

Originally characterized as angiogenic mediators, fibroblast growth factors (FGFs) exert autocrine and paracrine functions on tumor and stromal cells. FGFs play a pivotal role in the complex cross-talk among angiogenesis, inflammation, tumor growth, and drug resistance, thus contributing to tumor progression. FGF inhibitors may act as "two compartment" targeting drugs able to exert a deep impact on the growth of FGF-driven tumors. To date, the development of selective and non-selective tyrosine kinase FGF receptor inhibitors has been associated with significant toxicity. Recently, a different approach has been emerging, aimed at the development of extracellular "FGF ligand traps" able to bind and sequester FGFs, thus preventing their interaction with cognate receptors. This approach is based on the identification of natural FGF ligands followed by the development of small molecule mimetics endowed with a significant FGF binding/neutralizing capacity. Aim of this chapter is to provide an overview of the role of the FGF system in cancer and describe how the study of the FGF interactome has allowed the development of promising FGF-targeting molecules with potential implications for the therapy of FGF-driven tumors.

* University of Brescia, Department of Molecular and Translational Medicine, viale Europa 11, 25123 Brescia, Italy.

† Corresponding author: E-mail: roberto.ronca@unibs.it

1. Introduction

Fibroblast growth factors (FGFs) act as paracrine, autocrine or endocrine mediators in various physiological and pathological conditions. The FGF family encompasses 22 members, grouped into seven subfamilies on the basis of phylogenetic analysis and sequence homology.[1] The subfamilies FGF1/2/5, FGF3/4/6, FGF7/10/22, FGF8/17/18 and FGF9/16/20 act as canonical FGFs; FGF11/12/13/14 are intracellular factors acting in a FGF receptor (FGFR) independent manner; FGF19/21/23 subfamily members function as hormones (see Chapter 1 for details).[2]

Canonical FGFs are paracrine factors that trigger their biological responses by binding to and activating tyrosine kinase (TK) FGFRs. The interaction with heparan sulfate (HS) proteoglycans (HSPGs) plays a pivotal role in mediating the biological activity of FGFs, leading to the formation of signaling FGF/FGFR/HSPG ternary complexes.[3] Moreover, HSPGs sequester FGF molecules near the site of action, providing a reservoir for the growth factor and allowing the formation of extracellular matrix (ECM)-associated FGF gradients.[4] Due to the importance of HSPGs in the biology of FGF, HS and their structural analogue heparin have been the focus of numerous studies aimed at a better characterization of their role in the modulation of FGF activity and at the identification/development of putative anti-FGF drugs.

Intracellular FGFs act as signaling molecules in a FGFR-independent manner; they play a major role in neuronal functions at postnatal stages by interacting with intracellular domains of voltage-gate sodium channels and with the neuronal mitogen-activated protein kinase scaffold protein islet-brain-2.[5]

Hormone-like FGFs exhibit poor affinity for HSPGs, resulting in more diffusive properties through blood circulation.[6] These FGFs depend on Klotho co-receptors to activate intracellular signaling responses.[7] FGF19 (orthologue of murine FGF15) acts as a growth/differentiation factor in the heart and brain at embryonic stages and plays a crucial role in regulating hepatic bile acid production.[8] FGF21 is a metabolic regulator of lipolysis in the white adipose tissue[9] and FGF23 acts as a physiological regulator of phosphate and active vitamin-D blood levels.[10]

2. Aberrant Autocrine and Paracrine FGF Signaling in Tumors

An aberrant regulation of the FGF/FGFR system may occur in human tumors, leading to the deregulated activation of ligand-dependent or ligand-independent FGFR signaling. This may be the consequence of activating FGFR mutations that occur in the extracellular, transmembrane or TK domain of the receptor; chromosomal rearrangements that result in the expression of FGFR signaling fusion proteins; FGFR overexpression induced by gene amplification, translocation, aberrant transcriptional regulation or down-modulation of negative regulators; FGF overexpression by stromal and/or tumor cells, leading to the activation of autocrine/paracrine loops of stimulation (Fig. 1).

Even though most of the genomic aberrations discussed above lead to constitutive receptor activation and ligand-independent

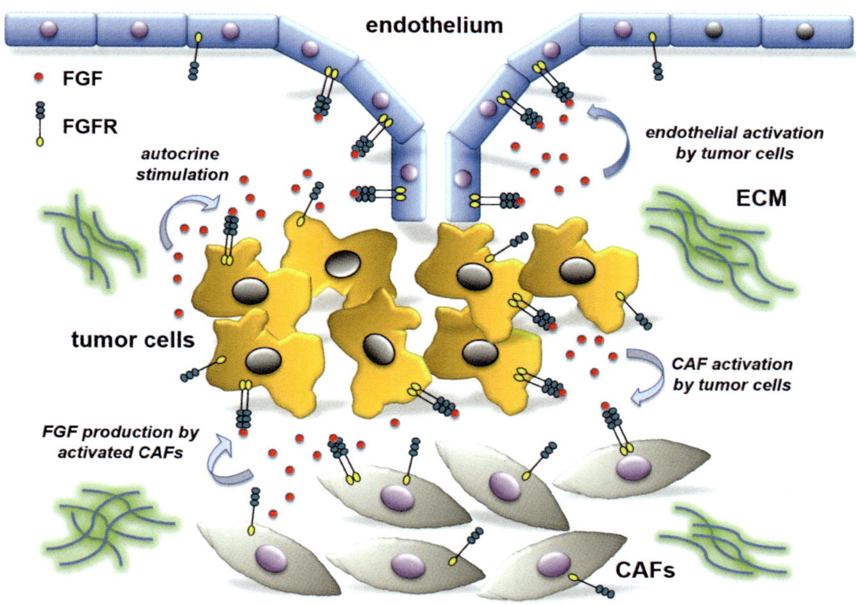

Fig. 1. FGF-mediated autocrine and paracrine loops of stimulation modulate the tumor/stroma cross-talk and angiogenesis. CAFs, cancer-associated fibroblasts.

signaling, the activation of ligand-dependent FGF/FGFR signaling appears to play an important role in the pathogenesis of cancer. This may occur via the activation of autocrine mechanisms of stimulation due to FGF production by cancer cells or may represent the consequence of the paracrine activity exerted on cancer cells by FGF(s) produced by the surrounding stroma (Fig. 1). In this context, several murine models have shown that ectopic FGF expression can promote cancer and that FGF overexpression by epithelial cells may induce carcinogenesis through an autocrine loop of stimulation. Examples include: FGF8 expression driven by the MMTV-LTR promoter that causes the occurrence of lobular-type mammary adenocarcinomas in mice at 1 year of age;[11] FGF8 expression in prostate epithelium that initiates prostatic intraepithelial neoplasia and prostatic cancer when occurs in a *Pten* haploinsufficient background;[12] the conditional expression of FGF10 in lung epithelium that induces pulmonary tumors.[13]

The first strong evidence for a role of autocrine FGF signaling in driving human tumorigenesis comes from seminal studies on melanomas that express high levels of FGFR1 and FGF2.[14] Since then, elevated levels of different members of the FGF family have been found in numerous human cancers.[15] Amplification of FGF1, resulting in increased FGF1 production, has been frequently observed in ovarian cancer and is associated with poor survival.[16] An aberrant autocrine FGF2/FGFR1-IIIc feedback loop of stimulation has been found in human non-small-cell lung cancer cell lines resistant to the epidermal growth factor receptor (EGFR) antagonist gefitinib.[17] Similar results were obtained for human head and neck squamous carcinoma cell lines. Indeed, FGF2 and FGFR co-expression frequently occurs in these cells, leading to an autocrine loop of stimulation that may involve also EGFR activation.[18] Several FGFs, including FGF1, FGF2, FGF5, FGF6, FGF7, FGF8, FGF9, FGF10, FGF17, FGF18 and FGF19 are upregulated in human prostate cancer[15] and murine studies have demonstrated the complex FGF/FGFR-dependent interplay between the epithelial and mesenchymal compartments in these tumors.[19]

Besides an aberrant activation of FGF-mediated autocrine loops of stimulation in tumor cells, paracrine modes of action of various FGFs may also contribute to tumorigenesis by acting on both parenchymal and stromal cancer cells. Increased plasma levels of FGF2 and other FGFs are found in multiple cancer types.[20] This partly reflects the increased release of FGFs as tumors invade and degrade the extracellular matrix,[21] free FGF molecules acting in turn as paracrine factors. Tumor cells may induce FGF2 release from the stromal inflammatory infiltrate,[22] thus promoting tumor cell survival via a paracrine loop of stimulation, and trigger a pro-angiogenic response in endothelial cells (see below). Neovascularization can be further augmented by an autocrine production of angiogenic FGF2 by endothelial cells.[22]

3. Angiogenic Activity of FGFs

FGFs are pleiotropic factors acting on different cell types, including endothelial cells. FGF1 and FGF2 represent the prototypical and best-studied pro-angiogenic members of the canonical FGF subfamily. *In vitro* they induce a complex "pro-angiogenic phenotype" in endothelial cells that recapitulates several aspects of the *in vivo* angiogenesis process, including modulation of endothelial cell proliferation, migration, protease production, integrin and cadherin receptor expression, and intercellular gap-junction communication (summarized in [23]). The angiogenic activity of FGF1 and FGF2 has been demonstrated in various *in vivo* experimental models, including the chick embryo chorioallantoic membrane (CAM),[24] rabbit/mouse cornea[25,26] and murine subcutaneous Matrigel plug[27] assays. However, FGF2 knockout and FGF1/FGF2 double-knockout mice develop normally even though they show a poor wound healing capacity when compared to control animals[28] whereas no abnormalities occur in mice lacking FGF1 only. The relatively mild phenotypic defects associated with FGF1/FGF2 deletion led to the hypothesis that the expression of other FGFs may exert a compensatory effect during development and under physiological conditions.

Besides FGF1 and FGF2, scattered pieces of information indicate that other FGFs are endowed with pro-angiogenic properties, whereas few or controversial data have been reported for other members of the family.[15] For instance, FGF1, FGF2, FGF4, FGF7 and FGF8b bind and activate FGFR1 or FGFR2, stimulating endothelial cell proliferation.[29,30] In addition, FGFs can modulate extracellular matrix degradation, as reported for the capacity of FGF1 and FGF2 to induce the secretion of MMP1 and MMP3 in endothelial cells[31,32] and the capacity of FGF2 to stimulate the shedding of endothelial membrane vesicles containing MMP1, MMP9 and metalloprotease inhibitors TIMP-1 and TIMP-2.[33] Also, various studies demonstrate that FGFs promote endothelial cell migration, as shown by the ability of FGF1, FGF2, FGF7, FGF16 and FGF18 to induce a chemotactic response in endothelium.[30,34–36]

4. The Role of FGFs in the Tumor/Stroma Cross-talk

Tumors are heterogeneous cellular entities composed of cancer cells and cells of the microenvironment in which they reside.[37] Stromal cells forming the tumor microenvironment include inflammatory cells (lymphocytes, macrophages, and mast cells), fibroblasts, and vascular components. The genetic basis of carcinogenesis implicates the acquisition of multiple genetic mutations in epithelial cells.[38] Then, tumor cells transform the surrounding stroma into a so-called "activated stroma"[39,40] that, in turn, can strongly influence/support tumorigenesis and tumor progression.[37] Thus, a reciprocal dynamic interaction occurs between tumor cells and activated stromal cells during cancer initiation and progression. This tumor-host communication interface mediates the proliferation of tumor cells at the primary site, the process of tumor angiogenesis, the migration and survival of cancer cells in the vasculature, and the growth of metastatic lesions at secondary sites through the autocrine/paracrine secretion of ECM proteins and growth factors.[41] Also, emerging evidences emphasize the ability of stromal cells to modulate tumor cell resistance or sensitization to different classes of therapeutics, depending on the specific microenvironmental context.[42] Thus, tumor microenvironment has become the focus of intense research, with the understanding that the alterations that occur in the tumor stroma might provide important prognostic hints, can affect the evaluation and selection of candidate drugs, and the generation of new therapeutic targets for various cancers.

As stated above, the FGF/FGFR system may play a critical role during carcinogenesis by regulating the cross-talk between epithelial and stromal tumor compartments. Enhanced FGFR signaling may have numerous effects on tumor progression, including promotion of cell proliferation, resistance to cell death, augmented motility and invasiveness, increased neovascularization, enhanced metastatic spreading and resistance to chemotherapy and radiation. Several studies have highlighted the importance of FGF/FGFR signaling in mediating epithelial-stromal interactions during prostate carcinogenesis.[43–46] For instance, overexpression of FGF10 in prostatic stroma by lentiviral delivery results in epithelial hyperproliferation that correlates with upregulation of androgen receptor expression.[19] Furthermore, the combination of FGF10 stromal overexpression with the epithelial expression of a constitutively activated form of Akt (myristoylated Akt1) results in cooperative effects on prostate tumorigenesis.[19] However, the translational significance of these murine models for human cancer remains unclear since FGF10 has not been found to be significantly expressed in human prostate cancer.[47] Nonetheless, it is conceivable that other members of the FGF family with receptor-binding specificities similar to FGF10 may be relevant in human prostate cancer, including FGF7 and FGF22. Also, activation of prostate tumor cell growth through androgen-independent stromal growth factor signals, such as FGF7, may occur under conditions of androgen deprivation.[48] These data may help to develop new therapeutic strategies to target the prostate tumor stroma under androgen-manipulated conditions. Interestingly, a recent study has demonstrated that downregulation of the micro-RNAs miR-15 and miR-16 in prostate cancer-associated fibroblasts (CAFs) promotes tumor growth and progression through the reduced post-transcriptional repression of FGF2 and of its receptor FGFR1.[49] Moreover, reconstitution of miR-15 and miR-16 significantly impaired the tumor-supportive capability of stromal cells *in vitro* and *in vivo*, thus enforcing the therapeutic concept aimed at reconstituting the expression of these micro-RNAs in advanced prostate cancer.[49]

Apart from its autocrine role in human melanoma, FGF2 may exert also paracrine functions in stroma formation during the progression of this tumor.[50] Indeed, FGF2 appears to act on fibroblasts and endothelial cells in order to modulate the tumor microenvironment, thus favoring melanoma growth, neovascularization, invasion, and metastasis.[14,50–52]

A recent study has identified FGF4 as a growth-promoting and radioprotective factor produced by CAFs in cervical cancer, leading to the activation of a tumor cell/CAF cross-talk that may confer a survival signal to overcome cell death in irradiated cervical cancer cells.[53] In addition, FGF2, FGF7 and FGF10 are implicated as autocrine and paracrine mediators of tumor-stroma interactions in pancreatic ductal adenocarcinomas.[54] In these tumors, mast cells, macrophages, and tumor cells overexpress vascular endothelial growth factor (VEGF)-A, VEGF-C, and FGF2 and this was highly correlated to intratumor microvessel density.[55]

Interestingly, the FGFR inhibitor PD173074 abrogates the rescue effect exerted by fibroblast supernatant on the cytostatic effects exerted by the TK inhibitor lapatinib on esophageal squamous-cell carcinoma cells.[56] These findings suggest a role for FGF/FGFR signaling in tumor drug resistance induced by stromal CAFs and suggest that a combination therapy with lapatinib *plus* a FGF/FGFR inhibitor might be effective in overcoming therapeutic resistance in esophageal squamous-cell carcinoma.

FGF2 is considered a potent angiogenic cytokine in multiple myeloma (MM). Both MM-derived cell lines and tumor cells isolated from the bone marrow of MM patients express and secrete FGF2, cell sorting studies indicating tumor cells as the predominant source of FGF2 in MM bone marrow.[57–60] Besides its pro-angiogenic functions, FGF2 plays also an important role in mediating tumor-stromal cell interactions in MM.[61] Indeed, bone marrow stromal cells (BMSCs) from MM patients express FGFR1-4 and stimulation of

BMSCs with FGF2 induces a time- and dose-dependent increase of interleukin-6 (IL-6), a potent growth/survival factor for MM cells.[62] Accordingly, IL-6 secretion is fully abrogated by anti-FGF2 antibodies, while stimulation with IL-6 enhances FGF2 expression by MM cell lines as well as by primary MM tumor cells, an effect inhibited by anti-IL-6 antibodies. These findings demonstrate a paracrine interaction between myeloma and bone marrow stromal cells triggered by the mutual stimulation of FGF2 and IL-6.

Finally, FGFs may activate a pro-inflammatory phenotype in endothelium,[63] indicating that the FGF/FGFR system may influence also the immune/infiltrate component of the tumor milieu.

All these considerations highlight the FGF/FGFR system as a critical player in tumor/stroma cross-talk in several cancer types. Thus, blocking the FGF/FGFR system may represent a "two-compartment" antitumor/antiangiogenic approach in cancer therapy.

5. Inhibition of the FGF/FGFR System by FGF Traps

Various approaches can be envisaged to neutralize the aberrant activation of the FGF/FGFR system that occurs in cancer, with its consequent effects on endothelial, parenchymal and stromal tumor compartments. In particular, we can distinguish between the possibility to prevent/modulate extracellular FGF-FGFR interactions or to impair intracellular signal transduction pathway(s) triggered by deregulated receptor activation.

Accordingly, the search for anti-FGF drugs currently under evaluation in cancer clinical trials (https://clinicaltrials.gov) indicates that two major classes of inhibitors of the FGF/FGFR system have been developed so far: FGFR selective and nonselective small-molecule TK inhibitors and anti-FGFR antibodies, a few studies focusing on decoy FGFR derivatives (see[64–69] for a detailed description of FGF/FGFR-targeting agents in phase I-III clinical development). Nevertheless, extracellular FGF ligands able to disrupt the interaction of the growth factor with its receptor (the so called FGF traps) may represent an interesting option for the treatment of

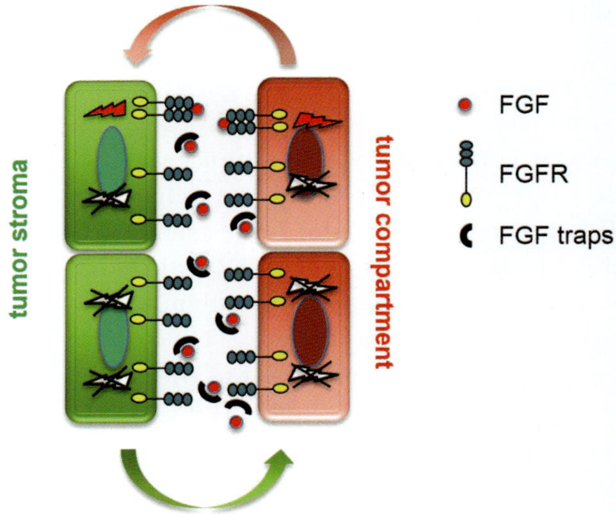

Fig. 2. FGF trap molecules sequester FGFR ligands, preventing receptor activation in tumor and stromal compartments.

tumors driven by an autocrine/paracrine loop of stimulation consequent to aberrant FGF expression (Fig. 2).

Based on the recognition that macromolecular interactions play a pivotal role in FGF biology, the study of FGF interactome in order to target key FGF interactions is gaining growing attention. However, targeting protein-protein contact interfaces may represent a challenging task. For instance, most of the surface contacts involve amino acids that are not contiguous in the primary protein sequence. Consequently, synthetic peptides derived from short primary amino acid sequences may represent a poor starting point for drug discovery. Nevertheless, various experimental evidences have been emerging that provide confidence in the search for FGF traps as antiangiogenic, anticancer compounds. Here, we will focus on the preclinical studies that have led to the identification of different classes of natural and synthetic FGF traps.

6. Identification and Development of FGF Traps

FGF traps are usually developed according to one of the following pipelines:

i) "random screening" of large libraries of peptides/antibodies,[70] glycans[71] and small chemical compounds[71] in the search for FGF binders endowed with inhibitory potential. This is usually performed by screening libraries of compounds as large as possible using appropriate biochemical or biological assays whose choice depends on a carefully balance among their biological relevance (i.e. the assay should mimic the biological process to be targeted), effortlessness, reproducibility and the possibility to be adaptable to high-throughput screening.[72,73] Also, a growing body of work points to the possibility to perform computer-based virtual screening procedures as a first step to "fish out" putative FGF-inhibiting structures to be subjected afterwards to biochemical/biological assays.[75]

ii) "rational design" based on known natural FGF-binding molecules (e.g. free or ECM-associated FGF-binding proteins[74] or glycans[75]). The process usually requires the identification of peptides or oligosaccharides representing a minimal FGF-binding sequence. Once this sequence has been identified, two main approaches can be undertaken to develop druggable FGF traps: the design of stable high-affinity natural/modified synthetic peptides/oligosaccharides or the development of mimetic small molecules. Then, the lead compounds can be refined by increasing their affinity for the targeted FGF(s) and their therapeutic potential by chemical modifications driven by molecular modeling and/or NMR or X-ray crystallographic data.

7. Decoy FGFR Derivatives

The "decoy" strategy takes advantage of truncated FGFR variants lacking the transmembrane and cytoplasmic domains. Interestingly, these free shortened FGFR variants are present in nature, bear the extracellular component of the canonical FGFR structures and maintain their native interaction properties for the FGF ligands. Recombinant and engineered extracellular-FGFR formats have been produced and characterized as efficacious FGF traps able to sequester FGF ligands and prevent activation of the FGF/FGFR signaling pathway. Truncated variants of FGFR1 IIIa (FGFR1 IIIa SR1)[76] and of FGFR2 IIIc (S252W FGFR2IIIc)[77] have been shown to bind to native ligands and prevent the physiological responses mediated by FGFs. The refinement of this strategy led to the production of FP-1039,[78] a soluble decoy receptor fusion protein composed of the full-length FGFR1 extracellular region and the human IgG1 Fc fragment. FP-1039 binds to all the mitogenic FGF ligands, inhibits FGF-stimulated cell proliferation *in vitro*, blocks FGF-induced angiogenesis *in vivo*, and inhibits *in vivo* growth of a broad range of tumor types. Of note, when compared to TK FGFR inhibitors, treatment with FP-1039 was not associated with toxicities or alterations in serum calcium and phosphate levels regulated by FGF-23.[78] To date, FP-1039 is the best-studied proteinaceous antiangiogenic FGF trap agent and is currently in phase II clinical trial.[65]

In order to further improve the therapeutic profile of FP-1039, an additional effort has been done to develop a series of soluble decoy receptor proteins by fusing various truncated extracellular regions of FGFR1 IIIc with the human IgG1 Fc fragment. One of these variants variant (named FGF-Trap) showed the highest affinity for FGF2, potently inhibited the FGF signaling pathway, suppressed FGF2-induced cell proliferation and migration, and reduced angiogenesis and tumor growth *in vivo*.[79]

8. Peptides and Antibodies

Phage display techniques have been used as a "fishing" approach to identify neutralizing peptides and antibodies that can bind to growth factor ligands. A few examples are available for the FGF family members.

A "mirror peptide" strategy was used to screen a phage display epitope library with anti-FGF2 antibodies. The "mirror" characteristic of the selected hexapeptides suggested structural similarity with FGFRs.[80] These hexapeptides were able to bind FGF2, prevented FGFR activation and suppressed basal and FGF2-induced proliferation of vascular endothelial cells at submicromolar concentrations.[80] Similarly, a phage display heptapeptide library was screened to identify 11 specific FGF2-binding clones and led to the identification of a heptapeptide (named P7) showing high homology and hydrophobic profiles similar to the immunoglobulin-like domain III (Ig-D3) of FGFR1 and FGFR2. Functional analysis demonstrated that P7 potently inhibits FGF-induced cell proliferation and neovascularization.[81]

Due to their role in tumor growth and angiogenesis (see above), FGF1 and FGF8b have been used as further "fishing baits" for the screening of phage display peptide libraries. Selection rounds on heparin-immobilized FGF1 allowed the identification of a 15-mer peptide with the conserved motifs SSG and VPS, corresponding to amino acid sequences 180–182 and 221–223 of the Ig- like domain II (Ig-D2) of FGFR1. The synthesized peptide inhibited the mitogenic activity of FGF1.[82] With a similar approach, 12 specific binding phage clones were selected against FGF8b. Among them, the heptapeptide P12 (HSQAAVP) showed high homology (and similar charges) to the Ig-D2 region of FGFR3 that directly participates in ligand binding. This peptide revealed a greater potential to prevent the binding of FGF8b to its receptor than other identified heptapeptides, caused a significant inhibition of FGF8b-induced cell proliferation and blocked the activation of $Erk_{1/2}$ and Akt signaling cascades in prostate cancer and vascular endothelial cells.[83]

Other peptides are in preclinical development even though in many cases their binding affinities are too low to support their therapeutic use. Nevertheless, these peptides can be used as templates for the design of peptidomimetics or improved peptide versions more effective in targeting specific cells or tissues when conjugated with therapeutic or diagnostic agents.[84,85]

Antibodies represent a promising class of therapeutics for the treatment of several pathologies, including cancer. FGFs have been used as antigens for wide libraries screening or immunization to fish out neutralizing antibodies able to bind with high affinity the growth factor and hampering FGFR interaction.

The screening of a human scFv phage display library led to the identification of a single chain Fv antibody (named 1A2) able to recognize FGF2 with high affinity and specificity, thus preventing its binding to FGFR. This scFv was then engineered as a full IgG antibody (named hIgG1-1A2) with high affinity for FGF2 ($K_d = 8 \times 10^{-9}$ M). *In vitro*, hIgG1-1A2 blocked various biological activities mediated by FGF2, including proliferation, migration and morphogenesis of human endothelial cells, and induced apoptosis of glioma cells.[86]

Among others, murine monoclonal antibodies have been generated against FGF2, FGF8b and FGF23 *via* classical *in vivo* immunization. The monoclonal antibody GAL-F2 represents one of the last examples of anti-FGF2 antibodies. GAL-F2 blocked the binding of FGF2 to all FGFRs, strongly inhibited FGF2-induced proliferation and downstream signaling in human endothelial cells, and inhibited proliferation, downstream signaling and *in vivo* growth of different hepatocellular carcinoma cell lines.[87] The anti-FGF8b antibody (KM 1334 clone) blocked androgen- and FGF8-stimulated growth of SC-3 mammary carcinoma cells.[88] Moreover, murine monoclonal antibodies against FGF23 (FN1, FC1, and FN2 antibodies) were developed by immunizing BALB/c mice with recombinant human FGF23. Their repeated administration increased serum phosphate and 1,25-dihydroxyvitmain D levels and enhanced mineralization of osteoid in adult *Hyp* mice.[89] Similarly, the derived humanized anti-FGF23 antibody KRN23 significantly increased the maximum renal tubular threshold for phosphate reabsorption (TmP/GFR), serum levels of phosphate and 1,25-dihydroxyvitmain D, and showed a favorable safety profile in X-linked hypophosphatemia (XLH) patients.[90] These data suggest caution about the use of non-selective inhibitors of the FGF/FGFR system that may cause undesired toxic effects due to the suppression of the activity of hormonal FGFs (see below).

9. Heparin/HS Derivatives

Cell-associated HSPGs act as FGF co-receptors whereas free HSPGs and structurally related heparin act as FGF antagonists by sequestering

Table 1: Heparin-like compounds that bind and inhibit different FGFs.

chemically modified heparins[120–125]	marine sulfated polymannuroguluronate[126]
heparin-derived oligosaccharides[127–129]	sulfated glycoconjugates[130]
sulfated beta-(1->4)-galacto oligosaccharides[131]	PI-88 and derivatives[132]
sulfated malto oligosaccharides[133]	linked sulfated tetracyclitols[134]
fucoidan[135]	Disulfated methyl 6-azido-6-deoxy-α-Dmannopyranosides[136]
pentosan polysulfate[137]	oversulfated exopolysaccharide from *Alteromonas infernos*[138]
sulfated K5 derivatives[139]	phosphorothioate oligodeoxynucleotides[140]
suleparoide (HS analogue)[141]	dextran derivatives[142]
β-cyclodextrin polysulfate[143]	dendrimer glucosamine conjugates[144]
carrageenan[145]	aurintricarboxylic acid[146]
HS mimetic M402[147]	TMPP (porphyrin analogue)[148]
synthetic HS[149]	suramin[150]
sucrose octasulfate[132]	suramin-like polysulfonated distamycine derivatives[90,130–133]
oligomannurarate sulfate JG3[151]	synthetic sulfonic acid polymers[98,143,152]

the growth factor in the extracellular environment, thus hampering their interaction with target cells. This identifies heparin and heparin-like molecules as putative ideal templates for the development of FGF traps.[75] However, the potent anticoagulant activity of heparin may limit its use as an antitumor drug. This has led to the development of a wide array of heparin-like anti-FGF molecules devoid of a significant anticoagulant activity, including chemically modified heparins, biotechnological heparins and non-saccharidic glycosaminoglycan mimetics (Table 1).

The prototypic molecule of this latter class of compounds is suramin, a polysulfonated napthylurea that exerts an interesting antiangiogenic activity directed against FGF2.[91] These observations prompted various efforts to develop naphthalenesulfonate derivatives endowed with a more specific antiangiogenic profile.[92–95] Another interesting class of synthetic non-saccharidic sulfated scaffolds is represented by sulfated flavonoids[96] whose antiangiogenic/anticancer potential has not been fully evaluated yet. Finally, sulfonic

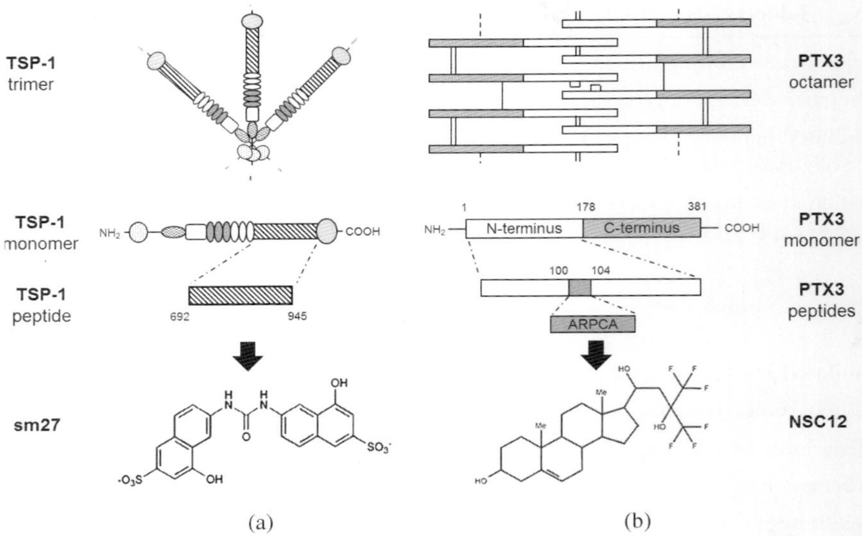

Fig. 3. Natural FGF-binding proteins as prototypes for the discovery of small molecule FGF traps. The figure schematizes the processes that led to the identification of the minimal FGF-binding regions in TSP-1 (a) and PTX3 (b) and of the corresponding small molecule peptidomimetics sm27 and NSC12.

acid polymers are organic acids that have the strong tendency to bind tightly to proteins and have been taken in consideration as antiangiogenic/antitumor compounds.[97,98]

A common feature of all these compounds is their scarce specificity due to their ability to interact with a variety of heparin-binding proteins, growth factors and cytokines.[96] Even though a multi-target effect may represent an interesting therapeutic opportunity, the safety profile of heparin-like derivatives should be evaluated carefully.

10. Thrombospondin-1 (TSP-1) and its Derivatives

TSP-1 is modular glycoprotein (Fig. 3A) secreted by different cell types that can be found as a circulating soluble molecule or as an ECM component. It is composed of multiple active domains that bind to cell receptors (e.g. CD36, integrins, HSPGs) and soluble factors.[99] In particular, TSP-1 binds to FGF2, preventing the accumulation of the growth factor in the ECM and generating inactive TSP-1/FGF2 complexes.[100] It also prevents the interaction of FGF2 with endothelial HSPGs and FGFRs, inhibiting FGF2 mitogenic and chemotactic activity in endothelial cells and exerting an antiangiogenic effect *in vivo*.[101] These observations suggest that free TSP-1 can act as a scavenger for matrix-associated FGFs, affecting their location, bioavailability and function, whereas ECM-associated TSP-1 may act as a "FGF sink", sequestering the growth factor in an inactive form.[99]

TSP-1/FGF2 interaction and its exploitation for the design of inhibitory peptide sequences represents a paradigmatic example of how the knowledge of naturally occurring protein-protein interactions can be exploited for the discovery of new drug-like inhibitors. At a molecular level, the interaction of TSP-1 with FGF2 is mediated by its COOH-terminal, antiangiogenic 140 kDa fragment. Starting from this evidence, the FGF2-binding sequence was identified by the analysis of a series of proteolytic fragments of TSP-1 and of smaller synthetic peptides representing key amino acid sequences of the protein.[100,101] Finally, peptide array technology combined with surface plasmon resonance analysis led to the identification of three potentially active linear amino acid sequences involved in the interaction of FGF2 with TSP-1.[102] On this basis, molecular dynamics and NMR analysis combined to a pharmacophore-based approach allowed the identification of the small TSP-1 mimetic molecule sm27 (Fig. 3A) [(4-hydroxy-6-[(8-hydroxy-6-sulfonaphthalen-2-yl) carbamoylamino] naphthalene-2-sulfonic acid)] able to bind FGF2 with a dissociation constant in the low μM range and to retain the antiangiogenic activity of the entire TSP-1.[102,103] NMR and MD data demonstrate that sm27 engages the heparin-binding site of FGF2 and induces long-range dynamics perturbations along FGF2/FGFR1 interface regions. Based on its chemical structure and its capacity to bind the heparin-binding site of FGF2, sm27 is anticipated to interact also with other heparin-binding proteins, as observed for heparin-like derivatives. Thus, its specificity and safety profile warrant further investigation. It is remarkable that different experimental approaches (*i.e.* TSP-1-based rational design and the screening of heparin-like naphthalene sulfonated derivatives[71]), led to the identification of structurally similar FGF traps.

11. Long Pentraxin-3 (PTX3) and Its Derivatives

The soluble pattern recognition receptor long pentraxin-3 (PTX3) is a member of the pentraxin family produced locally in response to inflammatory signals.[104] PTX3 may serve as a mechanism of amplification of inflammation and innate immunity with non-redundant functions in various physiopathological conditions including angiogenesis and cancer.[105] PTX3 shares the C-terminal pentraxin domain with short pentraxins and possesses a unique N-terminal domain[104] (Fig. 3B). The biological activity of PTX3 is related to its ability to interact with different ligands via its N-terminal or C-terminal domain as a consequence of the modular structure of the protein.[104,105] When assessed for the capacity to interact with a variety of extracellular signaling polypeptides, PTX3 was found to bind FGF2 with high affinity via its N-terminal extension,[106,107] thus inhibiting FGF2-dependent endothelial cell proliferation *in vitro* and angiogenesis *in vivo*.[105–108] Recently, it has been proven that PTX3 binds other FGFs via its N-terminal extension, including FGF6, FGF8b, FGF10 and FGF17.[46] Accordingly, transgenic PTX3 overexpression by tumor cells efficaciously impairs the activation and signaling of the FGF/FGFR system in FGF-driven tumor cell lines, thus affecting tumor growth and metastasis in different models of melanoma, prostate and mammary carcinomas.[46,109,110] Accordingly, PTX3 accumulation in tumor stroma, obtained through endothelial specific overexpression of PTX3 in transgenic mice, deeply hampers the tumorigenic, angiogenic and metastatic potential of various syngeneic FGF-dependent murine tumor cell lines.[111] Conversely, homozygous PTX3 inactivation in $PTX3^{-/-}$ mice enhances FGF-dependent angiogenesis, tumor growth and metastasis.[111]

A preliminary characterization of the binding of FGF2 to the PTX3 N-terminus demonstrated that the synthetic peptide PTX3(97-110) competes for FGF2/PTX3 interaction and, similar to full length PTX3 protein, it inhibits the angiogenic activity FGF2.[107]

Next, the acetylated pentapeptide Ac-ARPCA-NH2 (hereby named ARPCA), corresponding to the amino acid sequence 100-104 in the PTX3 molecule (Fig. 3B), was identified as the minimal FGF2-binding peptide able to interfere with the biological activity of the growth factor.[112] ARPCA appears to exert its antagonist activity by mimicking the hydrophobic FGF-binding region of Ig-D2 in FGFR, thus establishing hydrophobic interactions with the receptor-binding domain of FGF2 and competing with FGFRs for the binding to the growth factor.[112] Accordingly, ARPCA hampers the formation of a productive HSPG/FGF2/FGFR1 ternary complex and suppresses the angiogenic activity exerted by FGF2 in the chick embryo CAM assay and neovascularization driven by FGF2-overexpressing tumor cells in a zebrafish embryo/tumor graft assay.[112] In addition to FGF2, ARPCA also binds FGF8b, blocks its angiogenic and tumorigenic potential, and functions as a multi-FGF trap by impairing tumor burden of FGF2/FGF8b dependent tumors.[113]

In an effort to overcome the limitations due to the complex and proteinaceous structure of PTX3 and the limited stability of ARPCA peptide, a novel PTX3-derived small molecule FGF trap has been recently identified.[114] To this aim, the pharmacophore model of ARPCA/FGF2 interaction was used for *in silico* screening of small molecule databases, allowing the identification of NSC12 [4,4,4-trifluoro-1-(3-hydroxy-10,13-dimethyl-2,3,4,7,8,9,11,12,14,15,16,17-dodecahydro-1H-cyclopenta[a]phenanthren-17-yl)-3-(trifluoromethyl) butane-1,3-diol] as an ARPCA mimic.[111] NSC12 (Fig. 3B) binds FGF2, inhibits the formation of bioactive HSPG/FGF2/FGFR1 ternary complexes destabilizing the FGF2/FGFR interaction. Experimental evidences demonstrate that NSC12 interacts with all members of the canonical FGF subfamilies and inhibits the proliferation of various FGF-dependent murine and human cancer cell lines. Parenteral and gavage delivery of NSC12 inhibits FGFR activation, tumor growth, angiogenesis and metastasis in various FGF-dependent murine and human tumor models, with no effect on the blood levels of FGF23, calcium and phosphorus.[114] Notably, NSC12 does not affect the growth and vascularization of FGF-independent tumors. To date NSC12 represents the first FGF-specific low molecular weight FGF trap for the treatment of tumors in which the ligand-dependent activation of the FGFR pathway is an oncogenic driver.

12. Concluding Remarks

Even though FGFs were firstly characterized as angiogenic factors, they exert autocrine and paracrine functions in cancer by acting also on tumor cells and other stromal components. Thus, the FGF/FGFR pathway may represent a key player in tumor growth by regulating the complex cross-talk between stromal and tumor compartments. Accordingly, activation of the FGF/FGFR system by gene upregulation, oncogenic mutations or amplifications is implicated in key steps of tumor growth and progression.[2] Also, compensatory upregulation of the FGF/FGFR system may facilitate the escape from angiostatic anti-VEGF blockade.[115,116] Thus, experimental and clinical evidence provides a compelling biologic rationale for the development of anti-FGF/FGFR targeting agents in cancer therapy. In this scenario, the possibility to block the biological activity of FGFs by means of molecular traps that sequester the growth factor in the extracellular environment represents a potentially interesting therapeutic option for treatment of tumors in which the ligand-dependent activation of the FGFR pathway is an oncogenic driver.

FGF traps may exist as selective inhibitors (e.g. antibodies or peptides specifically directed against a given FGF) or nonselective inhibitors (e.g. heparin-like compounds that bind to different members of the FGF family as well as other heparin-binding growth factors). The nonselective binding would confer to these molecular traps the capacity to sequester multiple growth factors simultaneously, acting as 'multitarget agents'.[75] On the other hand, a broad binding capacity may cause nonselective FGF traps to exert undesired side-effects and/or toxicity. Similarly, multi-targeting TK inhibitors are endowed with toxicity profiles often related to their anti-VEGFR action, such as cardiovascular or hypertensive drawbacks or proteinuria, or with other side effects, like gastrointestinal disorders or skin reactions.[117] On the other hand, even though selective TK FGFR inhibitors show better tolerability in respect to non-selective TK inhibitors, they do exert relevant side effects (hyper-phosphatemia and tissue calcification) strictly correlated to the inhibition of the FGF23 pathway.[69,118] At variance, long-term administration of the small molecule FGF trap NSC12 does not affect the blood levels of phosphorus, calcium and FGF23 in tumor-bearing mice.[114] These findings are in keeping with the safety profile in murine tumor models of the FGFR1-derived FGF trap FP-1039[78] and of the allosteric multi-FGFR blocker SSR128129E.[119] These observations suggest that hyperphosphatemia may represent a side effect of TK FGFR inhibitors rather than of extracellular inhibitors of the FGF/FGFR system. Further studies are eagerly required to better assess the efficacy and safety profile of selective and nonselective FGF traps.

Acknowledgements

This work was supported in part by Associazione Italiana Ricerca sul Cancro (AIRC grant no. 14395) to M.P.

References

1. Itoh, N., and Ornitz, D.M. Functional evolutionary history of the mouse Fgf gene family. *Dev Dyn* **237**, 18–27 (2008).
2. Beenken, A., and Mohammadi, M. The FGF family: biology, pathophysiology and therapy. *Nat Rev Drug Discov* **8**, 235–253 (2009).
3. Richard, C., Liuzzo, J.P., and Moscatelli, D. Fibroblast growth factor-2 can mediate cell attachment by linking receptors and heparan sulfate proteoglycans on neighboring cells. *J Biol Chem* **270**, 24188–24196 (1995).
4. Hacker, U., Nybakken, K., and Perrimon, N. Heparan sulphate proteoglycans: the sweet side of development. *Nat Rev Mol Cell Biol* **6**, 530–541 (2005).
5. Goldfarb, M., *et al.* Fibroblast growth factor homologous factors control neuronal excitability through modulation of voltage-gated sodium channels. *Neuron* **55**, 449–463 (2007).
6. Kurosu, H., *et al.* Regulation of fibroblast growth factor-23 signaling by klotho. *J Biol Chem* **281**, 6120–6123 (2006).

7. Razzaque, M.S. The FGF23-Klotho axis: Endocrine regulation of phosphate homeostasis. *Nat Rev Endocrinol* **5**, 611–619 (2009).
8. Inagaki, T., *et al.* Fibroblast growth factor 15 functions as an enterohepatic signal to regulate bile acid homeostasis. *Cell Metab* **2**, 217–225 (2005).
9. Inagaki, T., *et al.* Endocrine regulation of the fasting response by PPARalpha-mediated induction of fibroblast growth factor 21. *Cell Metab* **5**, 415–425 (2007).
10. Itoh, N. Hormone-like (endocrine) Fgfs: their evolutionary history and roles in development, metabolism, and disease. *Cell Tissue Res* **342**, 1–11 (2010).
11. Daphna-Iken, D., *et al.* MMTV-Fgf8 transgenic mice develop mammary and salivary gland neoplasia and ovarian stromal hyperplasia. *Oncogene* **17**, 2711–2717 (1998).
12. Zhong, C., Saribekyan, G., Liao, C.P., Cohen, M.B., *et al.* Cooperation between FGF8b overexpression and PTEN deficiency in prostate tumorigenesis. *Cancer Res* **66**, 2188–2194 (2006).
13. Clark, J.C., *et al.* FGF-10 disrupts lung morphogenesis and causes pulmonary adenomas *in vivo. Am J Physiol Lung Cell Mol Physiol* **280**, L705–L715 (2001).
14. Wang, Y., and Becker, D. Antisense targeting of basic fibroblast growth factor and fibroblast growth factor receptor-1 in human melanomas blocks intratumoral angiogenesis and tumor growth. *Nat Med* **3**, 887–893 (1997).
15. Ronca, R., Giacomini, A., Rusnati, M., and Presta, M. The potential of fibroblast growth factor/fibroblast growth factor receptor signaling as a therapeutic target in tumor angiogenesis. *Expert Opin Ther Targets*, 1–17 (2015).
16. Birrer, M.J., *et al.* Whole genome oligonucleotide-based array comparative genomic hybridization analysis identified fibroblast growth factor 1 as a prognostic marker for advanced-stage serous ovarian adenocarcinomas. *J Clin Oncol* **25**, 2281–2287 (2007).
17. Marek, L., *et al.* Fibroblast growth factor (FGF) and FGF receptor-mediated autocrine signaling in non-small-cell lung cancer cells. *Mol Pharmacol* **75**, 196–207 (2009).
18. Marshall, M.E., *et al.* Fibroblast growth factor receptors are components of autocrine signaling networks in head and neck squamous cell carcinoma cells. *Clin Cancer Res* **17**, 5016–5025 (2011).

19. Memarzadeh, S., *et al.* Enhanced paracrine FGF10 expression promotes formation of multifocal prostate adenocarcinoma and an increase in epithelial androgen receptor. *Cancer Cell* **12**, 572–585 (2007).
20. Poon, R.T., Fan, S.T., and Wong, J. Clinical implications of circulating angiogenic factors in cancer patients. *J Clin Oncol* **19**, 1207–1225 (2001).
21. Ori, A., Wilkinson, M.C., and Fernig, D.G. The heparanome and regulation of cell function: structures, functions and challenges. *Front Biosci* **13**, 4309–4338 (2008).
22. Presta, M., *et al.* Fibroblast growth factor/fibroblast growth factor receptor system in angiogenesis. *Cytokine Growth Factor Rev* **16**, 159–178 (2005).
23. Javerzat, S., Auguste, P. and Bikfalvi, A. The role of fibroblast growth factors in vascular development. *Trends Mol Med* **8**, 483–489 (2002).
24. Ribatti, D., Vacca, A., Roncali, L., and Dammacco, F. The chick embryo chorioallantoic membrane as a model for in vivo research on anti-angiogenesis. *Curr Pharm Biotechnol* **1**, 73–82 (2000).
25. Herbert, J.M., Laplace, M.C., and Maffrand, J.P. Effect of heparin on the angiogenic potency of basic and acidic fibroblast growth factors in the rabbit cornea assay. *Int J Tissue React* **10**, 133–139 (1988).
26. Seghezzi, G., *et al.* Fibroblast growth factor-2 (FGF-2) induces vascular endothelial growth factor (VEGF) expression in the endothelial cells of forming capillaries: An autocrine mechanism contributing to angiogenesis. *J Cell Biol* **141**, 1659–1673 (1998).
27. Coltrini, D., *et al.* Matrigel plug assay: Evaluation of the angiogenic response by reverse transcription-quantitative PCR. *Angiogenesis* **16**, 469–477 (2013).
28. Miller, D.L., Ortega, S., Bashayan, O., Basch, R., *et al.* Compensation by fibroblast growth factor 1 (FGF1) does not account for the mild phenotypic defects observed in FGF2 null mice. *Mol Cell Biol* **20**, 2260–2268 (2000).
29. Presta, M., Rusnati, M., Urbinati, C., Statuto, M., *et al.* Functional domains of basic fibroblast growth factor. Possible role of Asp-Gly-Arg sequences in the mitogenic activity of bFGF. *Ann N Y Acad Sci* **638**, 361–368 (1991).
30. Mattila, M.M., *et al.* FGF-8b increases angiogenic capacity and tumor growth of androgen-regulated S115 breast cancer cells. *Oncogene* **20**, 2791–2804 (2001).

31. Pintucci, G., *et al.* Induction of stromelysin-1 (MMP-3) by fibroblast growth factor-2 (FGF-2) in FGF-2-/- microvascular endothelial cells requires prolonged activation of extracellular signal-regulated kinases-1 and -2 (ERK-1/2). *J Cell Biochem* **90**, 1015–1025 (2003).
32. Partridge, C.R., Hawker, J.R., Jr. and Forough, R. Overexpression of a secretory form of FGF-1 promotes MMP-1-mediated endothelial cell migration. *J Cell Biochem* **78**, 487–499 (2000).
33. Taraboletti, G., *et al.* Shedding of the matrix metalloproteinases MMP-2, MMP-9, and MT1-MMP as membrane vesicle-associated components by endothelial cells. *Am J Pathol* **160**, 673–680 (2002).
34. Terranova, V.P., *et al.* Human endothelial cells are chemotactic to endothelial cell growth factor and heparin. *J Cell Biol*, **101**, 2330–2334 (1985).
35. Gillis, P., *et al.* Keratinocyte growth factor induces angiogenesis and protects endothelial barrier function. *J Cell Sci* **112 (Pt 12)**, 2049–2057 (1999).
36. Antoine, M., *et al.* Fibroblast growth factor 16 and 18 are expressed in human cardiovascular tissues and induce on endothelial cells migration but not proliferation. *Biochem Biophys Res Commun* **346**, 224–233 (2006).
37. Polyak, K., Haviv, I., and Campbell, I.G. Co-evolution of tumor cells and their microenvironment. *Trends Genet* **25**, 30–38 (2009).
38. Vogelstein, B. and Kinzler, K.W. Cancer genes and the pathways they control. *Nat Med* **10**, 789–799 (2004).
39. Ronnov-Jessen, L., Petersen, O.W., and Bissell, M.J. Cellular changes involved in conversion of normal to malignant breast: Importance of the stromal reaction. *Physiol Rev* **76**, 69–125 (1996).
40. Tlsty, T.D., and Hein, P.W. Know thy neighbor: Stromal cells can contribute oncogenic signals. *Curr Opin Genet Dev* **11**, 54–59 (2001).
41. Pietras, K. and Ostman, A. Hallmarks of cancer: Interactions with the tumor stroma. *Exp Cell Res* **316**, 1324–1331 (2010).
42. McMillin, D.W., Negri, J.M., and Mitsiades, C.S. The role of tumour-stromal interactions in modifying drug response: Challenges and opportunities. *Nat Rev Drug Discov* **12**, 217–228 (2013).
43. Cunha, G.R., *et al.* The endocrinology and developmental biology of the prostate. *Endocr Rev* **8**, 338–362 (1987).
44. Hayward, S.W., *et al.* Malignant transformation in a nontumorigenic human prostatic epithelial cell line. *Cancer Res* **61**, 8135–8142 (2001).

45. Kwabi-Addo, B., Ozen, M., and Ittmann, M. The role of fibroblast growth factors and their receptors in prostate cancer. *Endocr Relat Cancer* **11**, 709–724 (2004).
46. Ronca, R., *et al.* Long pentraxin-3 as an epithelial-stromal fibroblast growth factor-targeting inhibitor in prostate cancer. *J Pathol* **230**, 228–238 (2013).
47. Ropiquet, F., Giri, D., Kwabi-Addo, B., Schmidt, K. *et al.* FGF-10 is expressed at low levels in the human prostate. *Prostate* **44**, 334–338 (2000).
48. Ishii, K., *et al.* Evidence that androgen-independent stromal growth factor signals promote androgen-insensitive prostate cancer cell growth *in vivo*. *Endocr Relat Cancer* **16**, 415–428 (2009).
49. Musumeci, M., *et al.* Control of tumor and microenvironment cross-talk by miR-15a and miR-16 in prostate cancer. *Oncogene* **30**, 4231–4242 (2011).
50. Lazar-Molnar, E., Hegyesi, H., Toth, S., and Falus, A. Autocrine and paracrine regulation by cytokines and growth factors in melanoma. *Cytokine* **12**, 547–554 (2000).
51. Meier, F., *et al.* Human melanoma progression in skin reconstructs: Biological significance of bFGF. *Am J Pathol* **156**, 193–200 (2000).
52. Ronca, R., *et al.* Long pentraxin-3 inhibits epithelial-mesenchymal transition in melanoma cells. *Mol Cancer Ther* **12**, 2760–2771 (2013).
53. Chu, T.Y., Yang, J.T., Huang, T.H., and Liu, H.W. Crosstalk with cancer-associated fibroblasts increases the growth and radiation survival of cervical cancer cells. *Radiat Res* **181**, 540–547 (2014).
54. Mahadevan, D., and Von Hoff, D.D. Tumor-stroma interactions in pancreatic ductal adenocarcinoma. *Mol Cancer Ther* **6**, 1186–1197 (2007).
55. Esposito, I., *et al.* Inflammatory cells contribute to the generation of an angiogenic phenotype in pancreatic ductal adenocarcinoma. *J Clin Pathol* **57**, 630–636 (2004).
56. Saito, S., *et al.* The role of HGF/MET and FGF/FGFR in fibroblast-derived growth stimulation and lapatinib-resistance of esophageal squamous cell carcinoma. *BMC Cancer* **15**, 82 (2015).
57. Vacca, A., *et al.* Bone marrow neovascularization, plasma cell angiogenic potential, and matrix metalloproteinase-2 secretion parallel progression of human multiple myeloma. *Blood* **93**, 3064–3073 (1999).
58. Otsuki, T., *et al.* Expression of fibroblast growth factor and FGF-receptor family genes in human myeloma cells, including lines possessing t(4;14)(q16.3;q32. 3) and FGFR3 translocation. *Int J Oncol* **15**, 1205–1212 (1999).

59. Di Raimondo, F., *et al.* Angiogenic factors in multiple myeloma: higher levels in bone marrow than in peripheral blood. *Haematologica* **85**, 800–805 (2000).
60. Sato, N., *et al.* Elevated level of plasma basic fibroblast growth factor in multiple myeloma correlates with increased disease activity. *Jpn J Cancer Res* **93**, 459–466 (2002).
61. Bisping, G., *et al.* Paracrine interactions of basic fibroblast growth factor and interleukin-6 in multiple myeloma. *Blood* **101**, 2775–2783 (2003).
62. Bommert, K., Bargou, R.C., and Stuhmer, T. Signalling and survival pathways in multiple myeloma. *Eur J Cancer* **42**, 1574–1580 (2006).
63. Presta, M., Andres, G., Leali, D., Dell'Era, P., *et al.* Inflammatory cells and chemokines sustain FGF2-induced angiogenesis. *Eur Cytokine Netw* **20**, 39–50 (2009).
64. Liang, G., Chen, G., Wei, X., Zhao, Y., *et al.* Small molecule inhibition of fibroblast growth factor receptors in cancer. *Cytokine Growth Factor Rev* **24**, 467–475 (2013).
65. Brooks, A.N., Kilgour, E. and Smith, P.D. Molecular pathways: fibroblast growth factor signaling: A new therapeutic opportunity in cancer. *Clin Cancer Res* **18**, 1855–1862 (2012).
66. Liang, G., Liu, Z., Wu, J., Cai, Y., *et al.* Anticancer molecules targeting fibroblast growth factor receptors. *Trends Pharmacol Sci* **33**, 531–541 (2012).
67. Katoh, M. and Nakagama, H. FGF receptors: Cancer biology and therapeutics. *Med Res Rev* **34**, 280–300 (2014).
68. Ho, H.K., Yeo, A.H., Kang, T.S. and Chua, B.T. Current strategies for inhibiting FGFR activities in clinical applications: Opportunities, challenges and toxicological considerations. *Drug Discov Today* **19**, 51–62 (2014).
69. Dieci, M.V., Arnedos, M., Andre, F. and Soria, J.C. Fibroblast growth factor receptor inhibitors as a cancer treatment: From a biologic rationale to medical perspectives. *Cancer Discov* **3**, 264–279 (2013).
70. Ronca, R., *et al.* Antiangiogenic activity of a neutralizing human single-chain antibody fragment against fibroblast growth factor receptor 1. *Mol Cancer Ther* **9**, 3244–3253 (2010).
71. Fernandez-Tornero, C., *et al.* Leads for development of new naphthalenesulfonate derivatives with enhanced antiangiogenic activity: Crystal structure of acidic fibroblast growth factor in complex with 5-amino-2-naphthalene sulfonate. *J Biol Chem* **278**, 21774–21781 (2003).

72. Rusnati, M., *et al.* Exploiting Surface Plasmon Resonance (SPR) Technology for the Identification of Fibroblast Growth Factor-2 (FGF2) Antagonists Endowed with Antiangiogenic Activity. *Sensors (Basel)* **9**, 6471–6503 (2009).
73. Rusnati, M. and Presta, M. Angiogenic growth factors interactome and drug discovery: The contribution of surface plasmon resonance. *Cytokine Growth Factor Rev* (2014).
74. Rusnati, M. and Presta, M. Angiogenic growth factors interactome and drug discovery: The contribution of surface plasmon resonance. *Cytokine Growth Factor Rev* **26**, 293–310 (2015).
75. Chiodelli, P., Bugatti, A., Urbinati, C. and Rusnati, M. Heparin/Heparan sulfate proteoglycans glycomic interactome in angiogenesis: Biological implications and therapeutical use. *Molecules* **20**, 6342–6388 (2015).
76. Guillonneau, X., *et al.* Fibroblast growth factor (FGF) soluble receptor 1 acts as a natural inhibitor of FGF2 neurotrophic activity during retinal degeneration. *Mol Biol Cell*, **9**, 2785–2802 (1998).
77. Tanimoto, Y., *et al.* A soluble form of fibroblast growth factor receptor 2 (FGFR2) with S252W mutation acts as an efficient inhibitor for the enhanced osteoblastic differentiation caused by FGFR2 activation in Apert syndrome. *J Biol Chem*, **279**, 45926–45934 (2004).
78. Harding, T.C., *et al.* Blockade of nonhormonal fibroblast growth factors by FP-1039 inhibits growth of multiple types of cancer. *Sci Transl Med* **5**, 178ra39 (2013).
79. Li, D., *et al.* A novel decoy receptor fusion protein for FGF-2 potently inhibits tumour growth. *Br J Cancer* **111**, 68–77 (2014).
80. Yayon, A., *et al.* Isolation of peptides that inhibit binding of basic fibroblast growth factor to its receptor from a random phage-epitope library. *Proc Natl Acad Sci USA*, **90**, 10643–10647 (1993).
81. Wu, X., *et al.* Isolation of a novel basic FGF-binding peptide with potent antiangiogenetic activity. *J Cell Mol Med* **14**, 351–356 (2010).
82. Fan, H., *et al.* Selection of peptide ligands that bind to acid fibroblast growth factor. *IUBMB Life* **49**, 545–548 (2000).
83. Wang, W., *et al.* Screening a phage display library for a novel FGF8b-binding peptide with anti-tumor effect on prostate cancer. *Exp Cell Res*, **319**, 1156–1164 (2013).
84. Yoo, M.K., *et al.* Targeted delivery of chitosan nanoparticles to Peyer's patch using M cell-homing peptide selected by phage display technique. *Biomaterials* **31**, 7738–7747 (2010).

85. Laakkonen, P., *et al.* Antitumor activity of a homing peptide that targets tumor lymphatics and tumor cells. *Proc Natl Acad Sci USA*, **101**, 9381–9386 (2004).
86. Tao, J., *et al.* Selection and characterization of a human neutralizing antibody to human fibroblast growth factor-2. *Biochem Biophys Res Commun*, **394**, 767–773 (2010).
87. Wang, L., *et al.* A novel monoclonal antibody to fibroblast growth factor 2 effectively inhibits growth of hepatocellular carcinoma xenografts. *Mol Cancer Ther* **11**, 864–872 (2012).
88. Tanaka, A., *et al.* High frequency of fibroblast growth factor (FGF) 8 expression in clinical prostate cancers and breast tissues, immunohistochemically demonstrated by a newly established neutralizing monoclonal antibody against FGF 8. *Cancer Res*, **58**, 2053–2056 (1998).
89. Aono, Y., *et al.* Anti-FGF-23 neutralizing antibodies ameliorate muscle weakness and decreased spontaneous movement of Hyp mice. *J Bone Miner Res*, **26**, 803–810 (2011).
90. Carpenter, T.O., *et al.* Randomized trial of the anti-FGF23 antibody KRN23 in X-linked hypophosphatemia. *J Clin Invest*, **124**, 1587–1597 (2014).
91. Takano, S., *et al.* Suramin, an anticancer and angiosuppressive agent, inhibits endothelial cell binding of basic fibroblast growth factor, migration, proliferation, and induction of urokinase-type plasminogen activator. *Cancer Res* **54**, 2654–2660 (1994).
92. Groen, H.J., *et al.* PNU-145156E, a novel angiogenesis inhibitor, in patients with solid tumors: A phase I and pharmacokinetic study. *Clin Cancer Res* **7**, 3928–3933 (2001).
93. Raman, K., Karuturi, R., Swarup, V.P., Desai, U.R., *et al.* Discovery of novel sulfonated small molecules that inhibit vascular tube formation. *Bioorg Med Chem Lett* **22**, 4467–4470 (2012).
94. Rusnati, M., and Urbinati, C. Polysulfated/sulfonated compounds for the development of drugs at the crossroad of viral infection and oncogenesis. *Curr Pharm Des* **15**, 2946–2957 (2009).
95. Urbinati, C., Chiodelli, P., and Rusnati, M. Polyanionic drugs and viral oncogenesis: A novel approach to control infection, tumor-associated inflammation and angiogenesis. *Molecules* **13**, 2758–2785 (2008).
96. Correia-da-Silva, M., Sousa, E., and Pinto, M.M. Emerging sulfated flavonoids and other polyphenols as drugs: Nature as an inspiration. *Med Res Rev* **34**, 223–279 (2014).

97. Bugatti, A., *et al.* Heparin-mimicking sulfonic acid polymers as multi-target inhibitors of human immunodeficiency virus type 1 Tat and gp120 proteins. *Antimicrob Agents Chemother* **51**, 2337–2345 (2007).
98. Liekens, S., *et al.* Modulation of fibroblast growth factor-2 receptor binding, signaling, and mitogenic activity by heparin-mimicking polysulfonated compounds. *Mol Pharmacol* **56**, 204–213 (1999).
99. Taraboletti, G., Rusnati, M., Ragona, L., and Colombo, G. Targeting tumor angiogenesis with TSP-1-based compounds: Rational design of antiangiogenic mimetics of endogenous inhibitors. *Oncotarget* **1**, 662–673 (2010).
100. Margosio, B., *et al.* Thrombospondin 1 as a scavenger for matrix-associated fibroblast growth factor 2. *Blood* **102**, 4399–4406 (2003).
101. Taraboletti, G., *et al.* The 140-kilodalton antiangiogenic fragment of thrombospondin-1 binds to basic fibroblast growth factor. *Cell Growth Differ* **8**, 471–479 (1997).
102. Colombo, G., *et al.* Non-peptidic thrombospondin-1 mimics as fibroblast growth factor-2 inhibitors: An integrated strategy for the development of new antiangiogenic compounds. *J Biol Chem* **285**, 8733–8742 (2010).
103. Pagano, K., *et al.* Direct and allosteric inhibition of the FGF2/HSPGs/FGFR1 ternary complex formation by an antiangiogenic, thrombospondin-1-mimic small molecule. *PLoS One* **7**, e36990 (2012).
104. Garlanda, C., Bottazzi, B., Bastone, A., and Mantovani, A. Pentraxins at the crossroads between innate immunity, inflammation, matrix deposition, and female fertility. *Annu Rev Immunol* **23**, 337–366 (2005).
105. Presta, M., Camozzi, M., Salvatori, G. and Rusnati, M. Role of the soluble pattern recognition receptor PTX3 in vascular biology. *J Cell Mol Med* **11**, 723–738 (2007).
106. Rusnati, M., *et al.* Selective recognition of fibroblast growth factor-2 by the long pentraxin PTX3 inhibits angiogenesis. *Blood* **104**, 92–99 (2004).
107. Camozzi, M., *et al.* Identification of an antiangiogenic FGF2-binding site in the N terminus of the soluble pattern recognition receptor PTX3. *J Biol Chem* **281**, 22605–22613 (2006).
108. Leali, D., *et al.* Fibroblast growth factor-2 antagonist and antiangiogenic activity of long-pentraxin 3-derived synthetic peptides. *Curr Pharm Des* **15**, 3577–3589 (2009).

109. Leali, D., *et al.* Long pentraxin-3 inhibits FGF8b-dependent angiogenesis and growth of steroid hormone-regulated tumors. *Mol Cancer Ther* **10**, 1600–1610 (2011).
110. Ronca, R., *et al.* Long Pentraxin-3 Inhibits Epithelial-Mesenchymal Transition in Melanoma Cells. *Mol Cancer Ther* (2013).
111. Ronca, R., *et al.* Long-Pentraxin 3 Derivative as a Small-Molecule FGF Trap for Cancer Therapy. *Cancer Cell* **28**, 225–239 (2015).
112. Leali, D., *et al.* Fibroblast growth factor 2-antagonist activity of a long-pentraxin 3-derived anti-angiogenic pentapeptide. *J Cell Mol Med* **14**, 2109–2121 (2010).
113. Giacomini, A., *et al.* A long pentraxin-3-derived pentapeptide for the therapy of FGF8b-driven steroid hormone-regulated cancers. *Oncotarget* **6**, 13790–13802 (2015).
114. Ronca, R., *et al.* Long-Pentraxin 3 Derivative as a Small-Molecule FGF Trap for Cancer Therapy. *Cancer Cell* **28**, 225–239 (2015).
115. Casanovas, O., Hicklin, D.J., Bergers, G., and Hanahan, D. Drug resistance by evasion of antiangiogenic targeting of VEGF signaling in late-stage pancreatic islet tumors. *Cancer Cell* **8**, 299–309 (2005).
116. Lieu, C., Heymach, J., Overman, M., Tran, H., *et al.* Beyond VEGF: inhibition of the fibroblast growth factor pathway and antiangiogenesis. *Clin Cancer Res* **17**, 6130–6139 (2011).
117. Sloan, B., and Scheinfeld, N.S. Pazopanib, a VEGF receptor tyrosine kinase inhibitor for cancer therapy. *Curr Opin Investig Drugs* **9**, 1324–1335 (2008).
118. Brown, A.P., Courtney, C.L., King, L.M., Groom, S.C., *et al.* Cartilage dysplasia and tissue mineralization in the rat following administration of a FGF receptor tyrosine kinase inhibitor. *Toxicol Pathol*, **33**, 449–455 (2005).
119. Bono, F., *et al.* Inhibition of tumor angiogenesis and growth by a small-molecule multi-FGF receptor blocker with allosteric properties. *Cancer Cell* **23**, 477–488 (2013).
120. Rusnati, M., and Presta, M. Interaction of angiogenic basic fibroblast growth factor with endothelial cell heparan sulfate proteoglycans. Biological implications in neovascularization. *Int J Clin Lab Res* **26**, 15–23 (1996).
121. Casu, B., *et al.* Short heparin sequences spaced by glycol-split uronate residues are antagonists of fibroblast growth factor 2 and angiogenesis inhibitors. *Biochemistry* **41**, 10519–10528 (2002).
122. Casu, B., *et al.* Undersulfated and glycol-split heparins endowed with antiangiogenic activity. *J Med Chem* **47**, 838–848 (2004).

123. Garg, H.G., *et al.* Anti-proliferative effects of O-acyl-low-molecular-weight heparin derivatives on bovine pulmonary artery smooth muscle cells. *Glycoconj J* **28**, 419–426 (2011).
124. Roy, S., *et al.* Bioactivity screening of partially desulfated low-molecular-weight heparins: A structure/activity relationship study. *Glycobiology* **21**, 1194–1205 (2011).
125. Kim, J., *et al.* Antiangiogenic and anticancer effect of an orally active low molecular weight heparin conjugates and its application to lung cancer chemoprevention. *J Control Release* (2014).
126. Wang, L., Geng, M., Li, J., Guan, H., *et al.* Studies of marine sulfated polymannuroguluronate on endothelial cell proliferation and endothelial immunity and related mechanisms. *J Pharmacol Sci* **92**, 367–373 (2003).
127. Liu, K., Yao, S., Lou, Y. and Xu, Y. Synthesis and anti-angiogenetic activity evaluation of N-(3-aryl acryloyl)aminosaccharide derivatives. *Carbohydr Res* **381**, 83–92 (2013).
128. Ashikari-Hada, S., Habuchi, H., Sugaya, N., Kobayashi, T., *et al.* Specific inhibition of FGF-2 signaling with 2-O-sulfated octasaccharides of heparan sulfate. *Glycobiology* **19**, 644–654 (2009).
129. Zhang, F., *et al.* Compositional analysis of heparin/heparan sulfate interacting with fibroblast growth factor.fibroblast growth factor receptor complexes. *Biochemistry* **48**, 8379–8386 (2009).
130. Liu, L., Ping Li, C., Cochran, S. and Ferro, V. Application of the four-component Ugi condensation for the preparation of sulfated glycoconjugate libraries. *Bioorg Med Chem Lett* **14**, 2221–2226 (2004).
131. Kasbauer, C.W., Paper, D.H. and Franz, G. Sulfated beta-(1→4)-galacto-oligosaccharides and their effect on angiogenesis. *Carbohydr Res* **330**, 427–430 (2001).
132. Cochran, S., Li, C.P. and Ferro, V. A surface plasmon resonance-based solution affinity assay for heparan sulfate-binding proteins. *Glycoconj J* **26**, 577–587 (2009).
133. Foxall, C., *et al.* Sulfated malto-oligosaccharides bind to basic FGF, inhibit endothelial cell proliferation, and disrupt endothelial cell tube formation. *J Cell Physiol* **168**, 657–667 (1996).
134. Cochran, S., Li, C.P. and Bytheway, I. An experimental and molecular-modeling study of the binding of linked sulfated tetracyclitols to FGF-1 and FGF-2. *Chembiochem* **6**, 1882–1890 (2005).
135. Giraux, J.L., *et al.* Modulation of human endothelial cell proliferation and migration by fucoidan and heparin. *Eur J Cell Biol* **77**, 352–359 (1998).

136. Liu, L., Li, C., Cochran, S., Jimmink, S., *et al.* Synthesis of a heparan sulfate mimetic library targeting FGF and VEGF via click chemistry on a monosaccharide template. *ChemMedChem* **7**, 1267–1275 (2012).
137. Zugmaier, G., Lippman, M.E. and Wellstein, A. Inhibition by pentosan polysulfate (PPS) of heparin-binding growth factors released from tumor cells and blockage by PPS of tumor growth in animals. *J Natl Cancer Inst* **84**, 1716–1724 (1992).
138. Matou, S., *et al.* Effect of an oversulfated exopolysaccharide on angiogenesis induced by fibroblast growth factor-2 or vascular endothelial growth factor *in vitro. Biochem Pharmacol* **69**, 751–759 (2005).
139. Presta, M., *et al.* Antiangiogenic activity of semisynthetic biotechnological heparins: Low-molecular-weight-sulfated Escherichia coli K5 polysaccharide derivatives as fibroblast growth factor antagonists. *Arterioscler Thromb Vasc Biol* **25**, 71–76 (2005).
140. Kitajima, I., Unoki, K. and Maruyama, I. Phosphorothioate oligodeoxynucleotides inhibit basic fibroblast growth factor-induced angiogenesis *in vitro* and *in vivo. Antisense Nucleic Acid Drug Dev* **9**, 233–239 (1999).
141. Benelli, U., *et al.* The heparan sulfate suleparoide inhibits rat corneal angiogenesis and *in vitro* neovascularization. *Exp Eye Res* **67**, 133–142 (1998).
142. Letourneur, D., *et al.* Heparin and non-heparin-like dextrans differentially modulate endothelial cell proliferation: *In vitro* evaluation with soluble and crosslinked polysaccharide matrices. *J Biomed Mater Res* **60**, 94–100 (2002).
143. Sakairi, N., Kuzuhara, H., Okamoto, T., and Yajima, M. Synthesis and biological evaluation of 2-amino-2-deoxy- and 6-amino-6-deoxycyclomaltoheptaose polysulfates as synergists for angiogenesis inhibition. *Bioorg Med Chem* **4**, 2187–2192 (1996).
144. Shaunak, S., *et al.* Polyvalent dendrimer glucosamine conjugates prevent scar tissue formation. *Nat Biotechnol* **22**, 977–984 (2004).
145. Hoffman, R., Burns, W.W., 3rd and Paper, D.H. Selective inhibition of cell proliferation and DNA synthesis by the polysulphated carbohydrate l-carrageenan. *Cancer Chemother Pharmacol* **36**, 325–334 (1995).
146. Gagliardi, A.R., and Collins, D.C. Inhibition of angiogenesis by aurintricarboxylic acid. *Anticancer Res* **14**, 475–479 (1994).
147. Zhou, H., *et al.* M402: A novel heparan sulfate mimetic, targets multiple pathways implicated in tumor progression and metastasis. *PLoS One* **6**, e21106 (2011).

148. Aviezer, D., *et al.* Porphyrin analogues as novel antagonists of fibroblast growth factor and vascular endothelial growth factor receptor binding that inhibit endothelial cell proliferation, tumor progression, and metastasis. *Cancer Res* **60**, 2973–2980 (2000).
149. Chen, J., *et al.* Enzymatic redesigning of biologically active heparan sulfate. *J Biol Chem* **280**, 42817–42825 (2005).
150. Kathir, K.M., Kumar, T.K., and Yu, C. Understanding the mechanism of the antimitogenic activity of suramin. *Biochemistry* **45**, 899–906 (2006).
151. Zhao, H., *et al.* Oligomannurarate sulfate, a novel heparanase inhibitor simultaneously targeting basic fibroblast growth factor, combats tumor angiogenesis and metastasis. *Cancer Res* **66**, 8779–8787 (2006).
152. Garcia-Fernandez, L., *et al.* Anti-angiogenic activity of heparin-like polysulfonated polymeric drugs in 3D human cell culture. *Biomaterials* **31**, 7863–7872 (2010).

Index